THE ICE PASSAGE

By the Same Author

Hail Mary Corner

Shadow of the Bear: Travels in Vanishing Wilderness

THE ICE PASSAGE

A TRUE STORY OF AMBITION,
DISASTER, AND ENDURANCE IN THE
ARCTIC WILDERNESS

BRIAN PAYTON

Library and Archives Canada Cataloguing in Publication
Payton, Brian, 1966–
The ice passage : a true story of ambition, disaster, and endurance in the arctic wilderness / Brian Payton.
ISBN 978-0-385-66532-2
1. Armstrong, Alexander, Sir, 1818-1899—Travel—Arctic regions.
2. Miertsching, Johann August, 1817-1875—Travel—Arctic regions.
3. Piers, Henry, 1818-1901—Travel—Arctic regions.
4. Investigator (Ship).
5. Arctic regions—Discovery and exploration—British.
6. Northwest Passage—Discovery and exploration—British.
7. Canada, Northern—Discovery and exploration—British.
8. Explorers—Great Britain—Biography.
9. Franklin, John, Sir, 1786-1847.
I. Title. G665.1850P39 2009 910.9163'27 C2009-902770-4

Printed and bound in the USA

Text design: Terri Nimmo
Maps: Paul Dotey

Published in Canada by Doubleday Canada,
a division of Random House of Canada Limited

Visit Random House of Canada Limited's website:
www.randomhouse.ca

10 9 8 7 6 5 4 3 2 1

FOR LILY

Table of Contents

ARCTIC NORTH AMERICA (Coastline Known to European Mariners in January 1850)

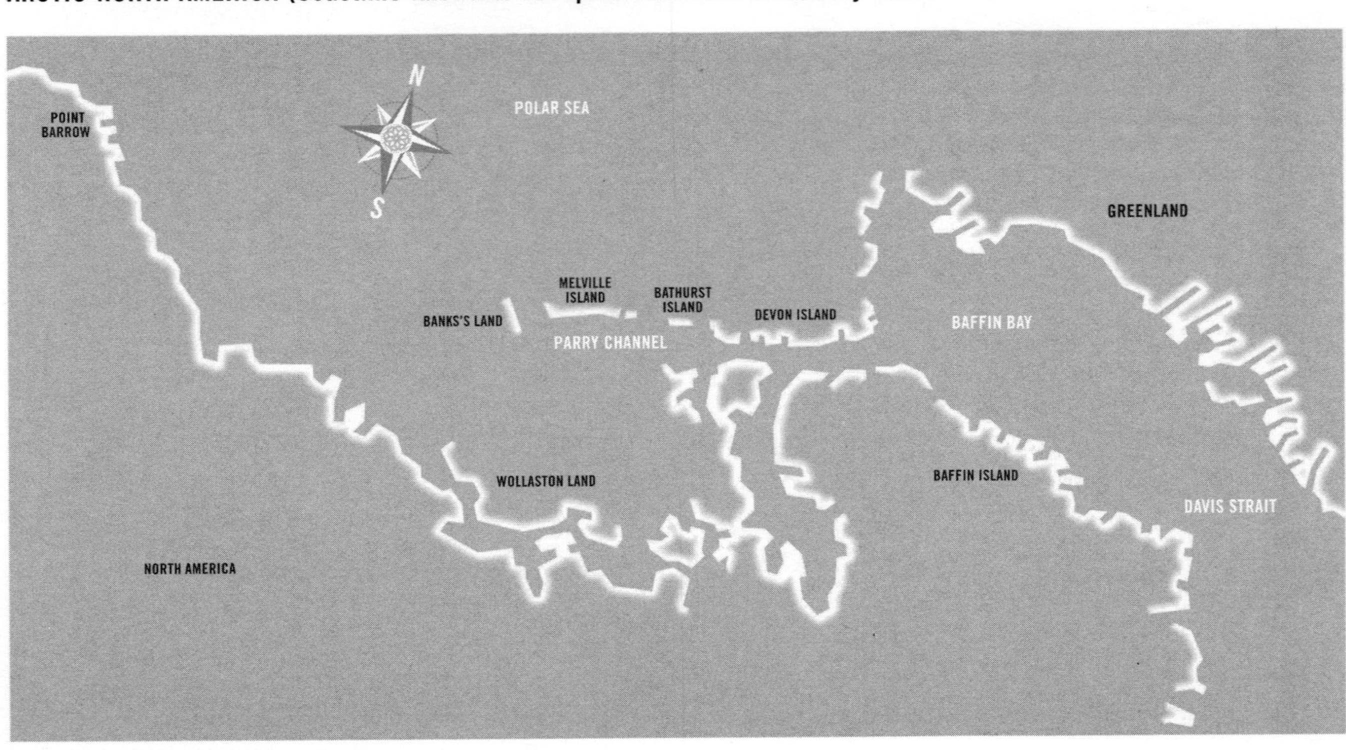

ARCTIC NORTH AMERICA (Journey of HMS *Investigator*)

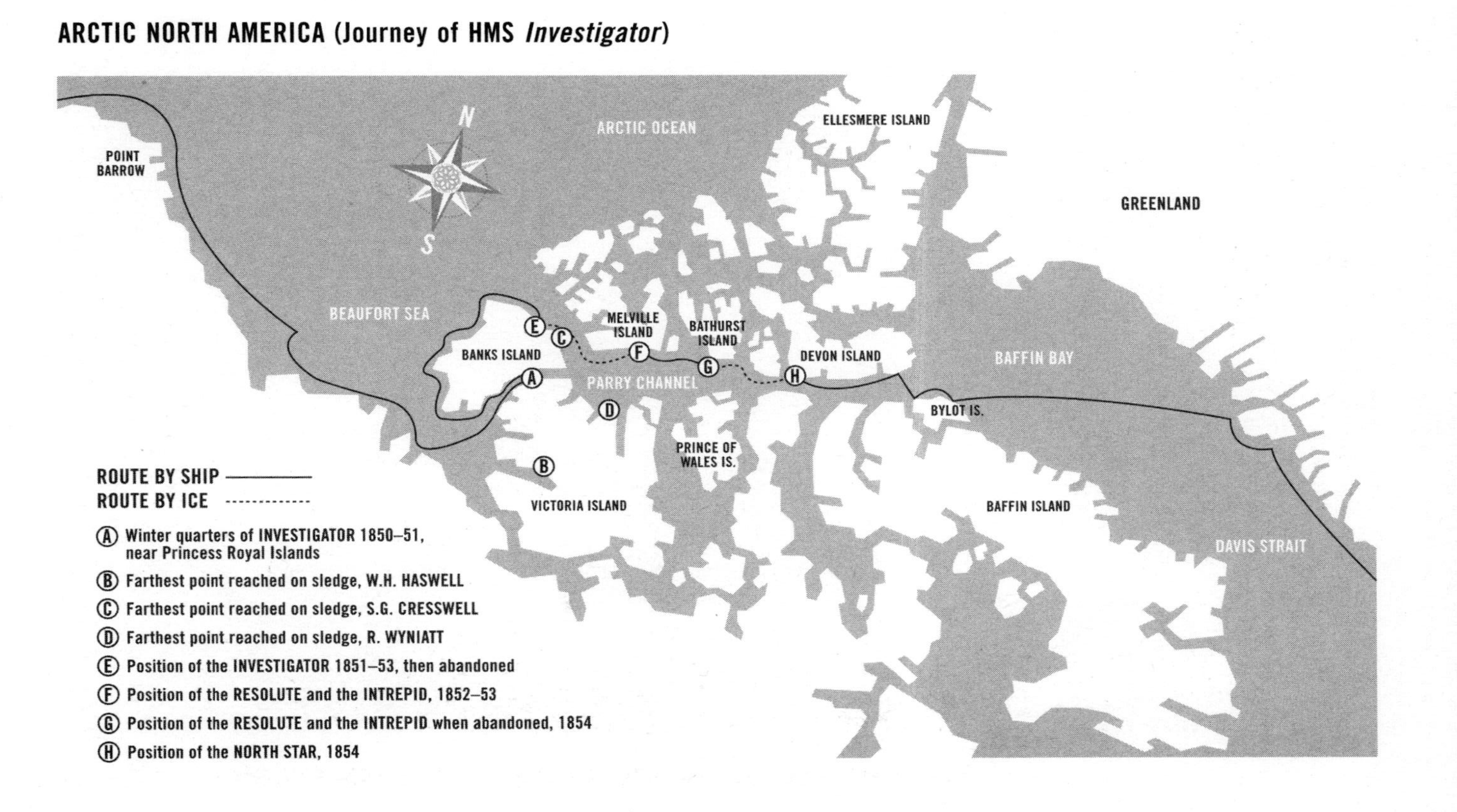

Prologue

RECKONING

ON SUNDAY, SEPTEMBER 16, 2007, a small craft sailed effortlessly around the Earth fourteen times, just as it had every day for nearly a dozen years. The payload module of RADARSAT-1 resembled an upright refrigerator swaddled in gold leaf. With its long, wing-like solar arrays stretched out on either side, it was a cubist's albatross on a perpetual glide above the sea. From five hundred miles above, it cast a cold and unblinking eye on great swaths of land, water, and ice. It surveyed the naked Earth through changing veils of cloud, snow, smoke, and smog. On this day, the images it captured and beamed back showed that a rift had opened at the top of the world.

The polar ice cap—a shifting, frozen mass the size of Europe—had been in steady retreat since the 1980s. In 2005, scientists were astounded by satellite images revealing the greatest single-year reduction in sea ice ever recorded. When the pictures arrived on September 16, 2007, they showed something much more ominous. That day turned out to be the year's "apparent Arctic sea ice minimum," the point at which summer melt comes to an end and autumn accumulation begins. It would also be, by far, the lowest extent of sea ice ever recorded. One million square miles of ice—an area nearly four times the size of Texas—had gone missing. For perhaps the first time in a million years, a large expanse of the Arctic Ocean was exposed, signalling both the realization of a long-held dream and a warning of potential catastrophe.

Polar ice scientists, oceanographers, and climatologists found themselves scrambling for new adjectives to render their sense of alarm and dread. However, having relayed dire warnings about polar ice retreat two years before, the media were slower in reporting this more profound event and its implications. This time around, leading the news in many parts of the world was the fact that the Northwest Passage—sea lane of myth and legend—was not only open for the first time, it was essentially ice-free.

We have wanted it badly and for a very long time, this lazy route between Europe and the Orient. Once found, it was to link the great markets and resources of the world, ushering in a new era of unprecedented wealth and power for those who could manage it. It was, after all, what inspired Christopher Columbus to sail boldly west to reach the East. Despite mounting contradictory evidence, the great navigator took to his grave the staunch belief that he had indeed reached the Indies and Cathay—that he had succeeded in making the dream come true.

Belief in this passage persisted as the map of the New World was gradually sketched in. The faith adapted, transformed, and evolved into three denominations: a Northwest Passage through North America, a Northeast Passage along the top of Eurasia, and—that ultimate prize—an Open Polar Sea. This least known and most improbable idea was tantalizing in the extreme. Proponents claimed that behind a seemingly impenetrable ring of ice lay a large, warm, open body of water at the pole.

Speculation about a temperate polar region stretches back to the earliest conception of a spherical Earth. While most ancient Greeks believed that the poles were frozen wastes, some claimed that beyond Thrace, where the god Boreas (the North Wind) dwelled, was Hyperborea—a land of leisure where the sun always shone. In early modern times, the belief in a navigable Polar Sea was first suggested in print in 1527 by Robert Thorne, agent of a prosperous Bristol-based trading company, whose father had been a crewmember of John Cabot's original 1496 voyage to Iceland. Thorne proposed the notion of an Open Polar Sea in a letter to Henry VIII. When the king showed little interest, Thorne published his hypothesis in a pamphlet entitled *Thorne's Plan*. In it, he declared: "there is no doubt, but sailing Northward and passing the Pole, descending to the Equinoctial line, we shall arrive at the Islands of Cathay, and it should be a much shorter way than any other." Explorers Willem Barents and Henry Hudson shared the belief. In the seventeenth century, England's Royal Society debated Dutch reports that they had actually discovered such a passage across the pole. The notion of an Open Polar Sea was eventually set aside, as neither scientific nor practical evidence could be found.

Then, in 1817, reports of an extraordinary breakup of sea ice occurred on the coast of Greenland and revived the faith. It would gain further credence as the search for the Northwest Passage gained momentum. The great Arctic explorer Sir William Edward Parry, a believer, continued the search for a way through the "barrier" of ice and into the open water. In 1845, Parry sent Sir John Franklin to continue the hunt for a Northwest Passage as well as for open water leading to the pole. Franklin's quest famously resulted in the disappearance of the two ships under his command, HMS *Erebus* and HMS *Terror*, as well as all 128 men in his expedition. As the years passed without word from Franklin, some suggested that he and his men could still be alive among an unknown race of people on the palm-fringed shores of the Open Polar Sea.

In an unprecedented international rescue effort, twenty-three ships were dispatched between 1848 and 1853 to find the lost explorer and solve, once and for all, the persistent mystery of the Northwest Passage. Although the tools, technology, and vessels employed were the finest available, the mariners' charts showed enormous gaps in knowledge. The price of filling those holes would ultimately be measured in human lives. One crew sent to the Arctic during that great search would succeed in being the first to circumnavigate the Americas and return with the answer to that larger, older question. They would find and cross the last link of the Northwest Passage and complete what was considered the greatest maritime achievement of the age.

On January 20, 1850, HMS *Investigator* set sail from Plymouth with a favourable wind. Her officers and crew would ultimately face starvation, madness, and death in an unknown wilderness. They would bring back the first sustained observations of the climate, flora, and fauna of the western Arctic Archipelago, and be among the first Europeans to contact the local native people. Theirs was a four-year battle with the polar ice—engine of oceanic currents, reflector of solar energy, regulator of the global atmosphere, habitat of the supremely adapted, testing ground of ambition and endurance, fortress of myths and dreams.

I must admit that this is not the story I set out to find. Like so many who have gone to explore or navigate some portion of the Arctic, I reached a point in my journey where I was obliged to change course. After a period of indecision, I abandoned all preconceptions and embarked in an unforeseen direction. I am very glad I did.

After months of preliminary research, I travelled to the western Arctic Archipelago in the summer of 2007 to collect information for a contemporary story about climate change. There, on Banks Island, I accompanied a veteran Inuit hunter as he tracked a herd of muskoxen. Along the way, he took the time to show me ancient stone caches and tent sites left behind by his ancestors, as well as newly parched marshes, melting permafrost, and vast tracts of coastline turned to mud and flowing into an almost iceless sea. He wanted me to observe for myself the transformation taking place in the Arctic. It is not the same world he knew as a boy—and the pace of change is accelerating. This is a message he has long been repeating to anyone willing to listen.

In the island's sole community (population 130), I interviewed elders—some of the last to be born in igloos and raised "on the land." They described the old ways of life, as well as dramatic and troubling changes observed on the ice, shore, and sea. I also learned about the connection between these families and the remains of a ship that had been abandoned on the north coast of the island at a place called Mercy Bay. It was a story that existed only in long-forgotten journals and passed-down memories.

That winter, I returned to Banks Island aboard an icebreaker, the Canadian Coast Guard Ship *Amundsen*. I was granted a berth among a large, international team of marine biologists, climatologists, and polar ice scientists—some of the foremost authorities in their fields. I accompanied them as they scooped, drilled, shaved, and shone light through the various stages of polar ice; as they surveyed the atmosphere above and measured the biomass hauled up from the water below. Among these researchers, I found a sense of urgency and purpose, as well as a growing unease. Like the Inuit who continue to rely on this environment, they found themselves duty bound to broadcast their increasingly dire news to the world.

Back home, sorting through all I'd learned, I was struck by how little we outsiders have come to understand about this immense and changing wilderness. Most of all, I was captivated by the evident link between our civilization's desires and its fate. For centuries, we have dreamed that the Arctic would turn out to be a different, more useful place than it appeared to be. From it, we wanted one thing above all else: a way through. We now seem certain to realize this vision, but at a cost both incalculable and unforeseen.

The final voyage of the *Investigator* is a profoundly revealing adventure. It is a story of ambition, pride, and ignorance that has come to haunt both land and sea. More than an elegy for imperiled wilderness, or forecast of drastic change to come—it reminds us why these things seem destined to be.

One

1.

DAYS OF PARADISE

July 1, 1850

He emerges as if from a shallow, anxious dream. After 150 days at sea, the brother bounds ashore on the island of Oahu. The air, redolent of tropical flora, a sultry 86 degrees. This stroll through Honolulu is his temporary release.

Johann August Miertsching, missionary of the Moravian Brotherhood, has had difficulty adjusting to his latest orders. At the urging of his superiors, the thirty-three-year-old native of Saxony committed to serving as interpreter for the Royal Navy aboard HMS *Enterprise* in the dual search for the lost Franklin expedition and the elusive Northwest Passage. Although he speaks several languages fluently, English is not yet among them. The voyage itself began with a fateful switch: he has not been travelling aboard the ship his superiors intended. Instead, the brother found himself temporarily assigned to HMS *Investigator*, consort of the *Enterprise*, amid rough and unruly men whose words he could scarcely comprehend. Since the *Investigator*'s departure from Plymouth on January 20, he has spent most of his days alone in his tiny cabin, devoting himself to the study of the English and Esquimaux languages, the discipline of private prayer, the avoidance of contact with the crew.

The brother is a robust and sturdy man whose broad, earnest face belies his fundamental empathy. His eyes are dark and intelligent; his hands, thick and workmanlike—ready to build the kingdom of God. His value to this expedition, however, is not as spiritual guide. He was chosen solely for his five years' experience among the natives at the Moravian mission on the coast of Labrador. There, he mastered the Esquimaux tongue, a skill in short supply. He undertook this study to make Christ known in one of the world's most desolate places. Whether this knowledge will serve as a foundation for communication with unknown Arctic peoples more than three thousand miles to the west is at best a matter of faith. Should contact be established, however, it might lead to the rescue of Franklin and his men and, eventually, the salvation of untold ignorant flocks. But these are challenges to be met in the coming months, in the distant Polar Sea. Now is the time to revel in a world of bright sensations: the fragrance of frangipani and sandalwood, the flavour of mango and coconut, the freedom and room to roam.

This liberty culminates in a pot of tea and the company of a local missionary. This Christian fellowship is long overdue. For Miertsching, the crude ways and words of British seamen give deep and frequent offence. One exception is the ship's assistant surgeon, Henry Piers, who travels with the brother this day. Less interested in devout conversation, Piers spends the afternoon in the garden of the mission house in pursuit of dazzling butterflies.

Back in the shade, Miertsching confirms the rumours he has heard. Although government is vested in King Kamehamea III and his ministers, it is the missionaries who wield true power. Honolulu is a city of over thirty thousand Europeans, Asians, and Polynesians. Here, the natives—once famous for seemingly boundless sexual freedom—are now clad in a reasonable facsimile of European dress and more or less submit to the evangelists. To the brother, it appears as a promising example of what can be attained through the moderating influence of religion. He is, however, aware of the missionaries' reputation. They are hated by many in this port city—especially his ship's officers and crew.

As a Christian who does not deny himself the occasional drink, Miertsching wonders if the zeal of local missionaries has overwhelmed their judgment. Their prohibition of every kind of alcoholic beverage is so extreme that they refuse to baptize the child of an otherwise upright man in the habit of taking a glass of wine at luncheon. Still, Miertsching listens politely to his host, basking in the glow of conviction and certainty. In two short weeks, the *Investigator* will again set sail and he will be met by those same impious faces for untold months to come. In the ice and desolation of the unexplored western Arctic, he will be left to keep the fire of faith for himself, his comrades, and any Esquimaux they might encounter. But to everything there is a time, and a time to every purpose. Now is the moment to breathe deeply, to gather in the fertile green, the warming sun, the call of tropical birds.

The passage from Plymouth to Honolulu was at once isolating and fraught with incident. As far as Brother Miertsching is concerned, these memories are best forgotten. But this will never be. For even as they shadow the present, they threaten to haunt the future.

Although he had previously made the crossing from Europe to Labrador and back again, the brother is no mariner. His lack of experience, his ignorance of English, his moral rectitude all serve to separate him from the officers and crew. He found his world reduced to a cabin of seven by seven feet—a berth, washstand, desk, and chair. It was there he confessed to his God and his journal that he felt acutely alone, trapped in a floating house full of strangers.

These strangers include fifty-seven British sailors, eight officers, and their captain—the ambitious Irishman Robert John Le Mesurier McClure. Their "house" is the 422-ton *Investigator*, an inelegant three-masted, copper-bottomed barque measuring 118 feet in length and 28 feet in breadth. Before departing on this, her second Arctic mission, the *Investigator* had recently been refurbished and provisioned for three full years. Rounded at both ends to aid in ice navigation, her hull is double-thick with English oak, Canadian elm, African teak. Still more timber and iron have been bolted over her bow and stern, resulting in upwards of twenty-nine solid inches. Stripped of ornament, she is painted a solemn coat of black.

Most of her officers and crew are hardened seamen caught up in the promise of a great and noble adventure. Many are specialists in their trades—from the ice master and surgeons, down to the caulker, blacksmith, and cook. A few have prior Arctic experience. Each officer has been assigned a servant to look after his washing, and Miertsching, a civilian, was granted this status. In order to converse with the Scottish seaman assigned to him, the brother found himself reduced to gestures. The few words of German to be found among the crew came from the lips of Able Seaman Charles Anderson, a Canadian Negro who had spent time working the immigrant ships between Europe and North America.

Although Miertsching's grip on English is tenuous, there was no mistaking the nature and tenor of the heated exchanges that flew amid the creaks and groans of the heavily laden ship. After five men were thrown overboard in a storm (but quickly retrieved), a bitter, open argument erupted between the officers and captain. The brother found this insolence remarkable.

Two weeks into the voyage, everything was soaked. Throughout the ship hung a powerful, fetid stench. To fight the damp, glowing hot cannonballs were placed in the living quarters, resulting in a fire. Although quickly extinguished, several sails were burned. The following night, the men threw the first of a series of raucous parties that aggravated the brother with their singing, dancing, and laughter. Occasionally, the officers would join in the revelry with the crew. In the privacy of his cabin, Miertsching would attempt to drown out the arguments and disorder, the thump and whine of the revellers, by playing his guitar and singing German hymns. Sundays offered the only respite. Then, the fiddle was set aside, the dancers took their rest, and the uproar would subside. Paradoxically, it was on the Lord's Day that he felt most alone.

Although a majority of the men aboard claim allegiance to the Church of England, the style of religion they practise on the high seas is far below the Moravian standard. The Moravian Brotherhood—the first global, large-scale Protestant missionary effort—espouses a life of self-denial and quick obedience. Moravians place great value on personal piety, the singing of hymns, and blessed unity. Unity is fostered at religious gatherings known as *lovefeasts*: services of Christian fraternity that seek to strengthen the bonds of goodwill and the forgiveness of past disputes.

The deteriorating order aboard the *Investigator* entered a new phase with the crossing of the equator and the attendant Festival of Neptune, Sovereign of the Sea. The pious brother had never seen nor heard of anything to compare. For him, it was a cause of serious offence. Those who had yet to "cross the line" were submitted to a humiliating hazing ritual in which men's bodies were smeared with tar—which was then scraped from their skin with a rusty barrel hoop. To Miertsching, the rough, bawdy nature of the spectacle seemed positively pagan, and the indignities to which these initiates were subjected horrific. In his prior work and experience, he had been confronted with godless and insolent men. But these were angels compared to the brazen sinners surrounding him aboard the *Investigator*.

The disputes between the officers and captain grew more frequent and sharper in tone. Incidents of insubordination increased. One day, four men were placed under arrest, with one singled out for four dozen lashes with the cat-o'-nine-tails, a whip of braided rope with nine thongs designed to inflict intense pain. A common punishment in the Royal Navy, the whippings were to ensure public humiliation and the ceremonial enforcement of authority. Some seamen have a cross tattooed on their backs to prevent them from being wantonly flogged. Although Miertsching deplores seeing a fellow human flayed, it seems the harsh rules of naval discipline are barely enough to keep them under control. In the first ten weeks since departing England, hardly a day passed without someone under arrest.

The voyage did provide moments of distraction, including occasional invitations to dine with the captain, a stroll in Patagonian hills, the view of the Southern Cross in unfamiliar skies. Miertsching enjoyed daily visits from Assistant Surgeon Piers, who tutored him in English. As there was little doctoring to do, Piers could afford time to indulge the isolated brother. A few hours of English instruction for the ship's interpreter could be crucial for their future success and survival.

On the Chilean coast, Miertsching narrowly missed transfer to the *Enterprise*. While the two ships lay at anchor together, the officers of the *Investigator* were invited to dine aboard the lead ship. There, Captain Richard Collinson, commander of the expedition, called the Esquimaux interpreter to gather his belongings and move them to his vessel. However, the night was wearing on, the whiskey flowing, and the weather growing grim. Captain Robert McClure requested that Miertsching be allowed to remain with him, aboard the *Investigator*, until landfall at the Sandwich Islands.

After a period of relative peace, a smouldering dispute between the officers and captain flared again. For Miertsching, it was beyond comprehension how educated and professional men could openly question their captain's authority on the high seas. How could they possibly get this ship safely through the South Pacific, let alone the Arctic ice? How could they hope to rescue Franklin and his men while at war with themselves?

As if triggered by the turmoil, sickness spread throughout the ship. Seventeen men fell ill with symptoms and conditions ranging from colds and rheumatism to "blood spitting" and colic. Miertsching was soon among them.

In this state, the men faced their first true test. The *Investigator* was struck at night by a squall. The sudden, sharp increase in wind—followed by an intense blast—damaged three topmasts and other critical rigging, resulting in the loss of a sail. They were foundering in the waves; everyone feared the ship would be lost. In an effort to regain command of his men and vessel, McClure flew into a rage. By morning, slackening winds allowed the crew to set sturdy yards in place of the broken mast. Control was finally regained.

At the first spare moment, McClure launched an inquiry, resulting in the trial of the officers and crew he believed responsible for losing control of the ship. The result was the arrest of First Lieutenant William Haswell. McClure promised further charges would be laid as soon as they reached the Sandwich Islands.

For Miertsching, that day could not dawn soon enough. On that day, he would trade the insolence and mayhem of the *Investigator* for the imagined peace and order of the *Enterprise*, the ship to which he was originally promised, the ship where superior accommodation, and perhaps more congenial company, awaited.

Then, as if in answer to prayer, the officers and crew rallied. With available tools and resources, they repaired much of the damage, painted, cleaned, and put the *Investigator* back in working order. The weather improved—along with the spirits of the crew. Drinking water was rationed to a single pint per day, but an extra measure of grog (rum and water) was frequently dispensed, further elevating the mood. The singing, dancing, and merriment that had marked the first weeks of the voyage resumed, but now the brother took less offence. His health returned. He emerged from his cabin more often and began distributing tracts on Christian values and views.

Miertsching was again invited to dine with the captain. In a ship plagued by insolence and insubordination, McClure had come to view the brother as impartial company. The brother—in a test of his English proficiency—steered the conversation on a spiritual course. He explained that he had been proselytizing among the men. At this, McClure laughed loud and long. British seamen, he assured his guest, are unlike the simple Labrador Esquimaux. They will not be so easily swayed.

At last, on Sunday, June 30, the Investigators sighted the islands of Hawaii, Maui, and Oahu. Both officers and crew lined the gunwale that day, held captive by the sight of those magnificent and storied isles, hungrily imagining the pleasures awaiting there. It seemed a paradise in which to loosen one's shirt and belt, a place to wallow at the warm belly of the world before turning north to face the ice. The brother, too, was moved by the lush spectacle before them. He felt drawn, not by baser instincts, but by the sound of church bells pealing and the sight of dark-skinned natives, dressed in white, on their way to worship.

The clear and brilliant sky rains down taxing heat. Regardless, Miertsching takes a carriage tour of the surrounding countryside to exercise his liberty. He revels in the abundance: lowlands crowded with breadfruit and banana, orderly plantations of familiar European fruits and vegetables arrayed on the hillsides above. It is, he observes, a verdant land of extraordinary beauty. As the brother takes in the sights with local religious leaders, the rest of the *Investigator*'s officers and crew are scattered across the city and beach below, engaged in a wholly different sort of leave.

With a determination fuelled by over four months at sea, the men quickly set about violating the numerous prohibitions declared by local missionaries. They have heard rumours that pale-skinned men are favoured by native ladies. First Mate Hubert Sainsbury, a particularly fair member of the ship's company, promptly tests the theory. He sits down next to a fetching native girl spied along the side of the road. Much to the amazement of his comrades, his face is soon being stroked adoringly.

While some men find their way straight into the arms of women, others are determined to get to the bottom of a cask. Many are incapacitated with drink; five will soon require treatment for venereal disease. Such behaviour is officially discouraged but wholly expected of men who may be without for several years to come. One group of seamen get their hands on a team of horses, which they ride back and forth along the sugar sand beach at a gallop, raising hell en route. Caught up in the exuberance of freedom, they ride and run without boots, cutting their tender feet. Half a dozen men recreate with such intensity that they collapse—requiring several days' recovery.

Out in the harbour, within the humid confines of the ship, Captain McClure is in an anxious state. Upon arriving in Honolulu, he discovered that his consort, the *Enterprise*, had left the day before his arrival. If the *Enterprise* reaches the ice first, McClure believes, Collinson will press on—and succeed—without him.

He turns his attention to the prosecution of First Lieutenant Haswell in a five-hour court of inquiry. The ship is bound for uncharted Arctic waters; insolence and insubordination will not be tolerated. After much debate, McClure is reluctantly persuaded to reinstate the accused after extracting a public pledge of obedience from him and his fellow officers. The bitter, unassailable truth is that he has no choice. He needs these men—each and every one.

Miertsching, enjoying yet another guided carriage tour, is impressed with the rapid growth of the city. Its harbour has anchorage for two hundred ships, some of which bear prefabricated houses of wood and steel from both England and the United States. These structures pop up on shore at a rate of one a day. He also notes the rising contingent of Americans on this island, which has long been under British influence. So large has their presence grown that today, July 4—U.S. Independence Day—is celebrated openly. At noon, he attends one such party at a local seminary. There, he is handed a letter from Captain McClure. He unfolds the note and discovers that all hands are ordered to return to the ship immediately, no later than four o'clock this afternoon. The *Investigator* sails today.

The brother looks up at the welcoming faces of his hosts, perturbed by this news. The original plan was to rest at the Sandwich Islands for fourteen days—not four. Still, it is not his to question. These are to be his last precious hours of freedom.

Following the banquet, with its polite conversation and tart lemonade, he finds himself surrounded by missionaries at an impromptu *lovefeast* in honour of his parting. His temporary, spiritual family sings songs of praise and offers prayers for a safe voyage. As the brother savours one last hymn, he can't help but think of the rude company into which he is about to return, perhaps for several years. He reflects on the loneliness and isolation ahead and feels pain at this parting. It is a pain born not of fear, but resignation. He resolves to endure the isolation, danger, and possible death for a greater glory. In the saturated hues of the tropical afternoon, in the warm embrace of newfound friends, the brother receives gifts and best wishes for success in the distant Polar Sea.

The walk from the seminary to the beach is far too short. He is escorted by two brethren who, in a final gesture, place watermelons in his arms. He steps aboard the waiting boat, then slips into the sparkling harbour. Along the way, he says a small prayer to God and his angels for guidance and protection.

As his boat approaches the *Investigator*, he watches the crew busily prepare the ship for departure. They whistle, sing, and curse as they go about their tasks—the resulting cloud of profanity assaults his sensibilities. Once onboard, he reports to the tight-lipped captain who, in a bid to regain his composure, invites the brother for a private drink.

At forty-two, Robert McClure's hairline is in full retreat. Along the sides and back of his head, he lets grow the remaining fringe, which he combs forward at the temples. He wears a trim neck beard in the current fashion. Also known as a "scarf," it is often grown to hide a double chin.

On McClure, however, it frames deep blue eyes, ruddy cheeks, and an expression both inquisitive and daring. Now, in early middle age, he has lived long enough to understand that men are offered few if any chances at greatness. The fact that his opportunity can only be seized though collaboration—by relying on others—is nothing short of maddening.

After a calming glass of wine, the captain and the brother proceed to the upper deck for roll call. There, it is revealed that three men have jumped ship, with all their belongings. A group of men who had been thrown in jail on shore have only just returned, their release secured by McClure's willingness to pay their fines. Five sick men have been transferred to another vessel for passage home; five new sailors have signed on to replace them.

By five o'clock in the afternoon, the *Investigator* clears Honolulu Harbour and sails in the open ocean. A strong wind pushes hard, and the Sandwich Islands, awash in amber light, rapidly shrink from view.

As night falls, the brother retires to his tiny cabin. He closes the door, reaches for his guitar, and sings hymns of hope and courage.

2.

THE WHITE EDGE

The wind is strong and fair, the air sharp and bright. At six o'clock, the morning peace is shattered by a loud cry from the crow's nest. The brother joins the scramble up top for a view as the fragments advance across the sea to meet them. Beneath an unmitigated sun, the bright white spreads and grows with fearsome grace and beauty. For most men crowding on deck, it is a spectacle unlike anything they have previously seen. It is what they have been waiting for. The charged, dizzying sensation of at last meeting the foe is like having climbed to the edge of a precipice over which they must now will themselves to jump.

The temperature has plunged over the four-week voyage north through the Aleutian Islands, the Bering Sea, and around the northwest tip of the continent. Now, on August 2, it falls further still, to 38 degrees. As if the ice has been lying in wait, it suddenly rushes south to meet them. Within two short hours, the *Investigator* is encircled in flat masses of ice, and yet the ship sails further north. The floes show between six and eight feet of freeboard, with a draft of eighteen to twenty-four feet hidden below the waterline. The sudden shocks, concussion, and scrapes against the hull test the nerve of the crew, now well beyond the sight of land. After the first wave sails past, the density increases, and then what appears to be solid pack approaches. To the crew's surprise, the sea remains navigable—with care. Could they press further still?

By two o'clock, the ship is trapped and motionless. From the crow's nest, word comes down that, from one end of the horizon to the other, the ice is now complete. Skill and nerve are required to bring the ship about and make for safer water.

Along the way, the men encounter the first inhabitants of this floating archipelago, lying on the floes in huddles of a hundred or more, basking in the twenty-four-hour sun. Few of the Investigators have seen or even heard tell of such beasts. Some make wild guesses as to what class of creature they may be; most stare in rapt wonder at the grunting and growling multitudes.

The larger males stretch to twelve feet in length and weigh perhaps a ton and a half. When they slip down into the water, the floes they leave behind rise by as much as two feet. In the water, females are encountered with pups at their side, which crawl up onto their mother's back at the approach of the ship. All have round heads, small eyes, and no visible ears. Their thick skin, gathered in deep folds at the shoulders, is covered with short, reddish hair, making them appear as if they've rolled in cinnamon powder. Their muzzles are short and broad, sporting prodigious moustaches of stiff, quill-like whiskers. Both males and females possess imposing ivory tusks, some as long as a man's leg. They come and go from the edge of the ice, diving to feed on molluscs in the shallows, hauling out to take their rest. Like walrus populations known to exist in the Atlantic Ocean and Laptev Sea, these creatures spend a large portion of their lives atop Arctic ice, relying on it as a platform for both hunting and giving birth.

Aboard the *Investigator*, some of the more experienced seamen observe that these "marine beef" have a reputation for yielding delicious steaks. A rifle is loaded and a ball fired into a near specimen. However, wishing no further delay, the captain calls off the attack.

Sailing through the low, shifting maze, the crew seeks retreat to the southeast. But the dark water cracks are squeezed tight between the floes until the sea itself is a vast, stained-glass window of winter. When the ice presses in to trap them, it becomes necessary to dispatch a team to the surface for assistance. The men compete for the right to be the first to reach it. To the boatswain goes the honour. After finding his balance, he ceremoniously reaches down and touches the hard Polar Sea.

The men stow their summer clothes and gear. For a second day, the crew deftly tacks to and fro in the hope of avoiding collisions. As the brother watches, he recalls the ice he saw streaming down from Baffin Bay and Davis Strait on the far side of the continent. Those often immense formations dwarf the largest ships and are made of ancient fresh water suspended in time—up to fifteen thousand years. True icebergs, born on land from calving glaciers, are not seen in these western waters. Although less dramatic, the sea-ice floes encountered here are no less a danger. The men who have no ice experience of any kind find their previous notions at odds with the surrounding reality.

"Had I known that the ice was so hard and strong," one seaman is heard to say, "I would have been only too glad to stay at home."

The ice now striking the *Investigator*'s hull is far younger than anyone imagines. This multi-year pack ice begins as autumn frazil, fine needle-like crystals suspended in water. This forms grease ice, a thin, slushy frazil layer so named for its oily appearance on the surface of the sea. If conditions are relatively calm, this slush congeals into fields of soggy snowballs or schools of small, translucent ice jellyfish hanging just below the surface. These in turn become fragile rafts, and then finally solid *cakes* (up to twenty-two yards across) and *floes* (greater than twenty-two yards) that eventually freeze together until the sea itself is at last locked away.

As the ice solidifies, the crystals relentlessly expel brine down to the water below, evolving and densifying into hard, clear ice that is essentially salt-free. Some will anchor to coves and bays as land-fast ice; most will become part of the contiguous polar ice cap. In all, Arctic sea ice survives no more than nineteen years before finding itself expelled to the edge of the pack. There, it is broken and cast adrift as the entire mass slowly rotates clockwise like a dense, frozen cyclone afloat at the top of the world. To the captain, officers, and crew of the *Investigator*, the ice they encounter here could well be the "barrier ice" separating the known world from the Open Polar Sea.

The brother's experience on the coast of Labrador provides little comfort. Everything about the ice seems designed to warn them off, to force them to turn their ship and flee or find themselves forced over and down. On only their second day amid the jostling cakes and floes, Miertsching wonders how much longer they will have to endure the perilous ice before finally reaching Greenland, some twenty-five hundred nautical miles to the east. The thought moves him to sing a few verses of a Moravian traveller's hymn:

Lord, speed our vessel in its course:
Let wind and waves propitious be;
Let Thy divine protection shield
All whom we now commend to Thee.

Three days on, the *Investigator* is held motionless, caught in the grip of the ice. Neither the captain nor the officers are sure of their location. Soundings are taken.

Like many units of measurement, the fathom was originally derived from the dimensions of the human form, in this case the distance between the middle fingertips of the outstretched arms of an ideally proportioned man. It was, literally, as far as he could reach. Over the centuries, this intimate way of measuring our world became standardized to represent a thousandth of a British nautical mile, or six feet. Here, despite soundings of a mere five to seven fathoms, there is no land in sight. Clouds and fog descend.

Since departing the Sandwich Islands, relations between the officers and captain have markedly improved. The antagonism that culminated in the inquiry in Honolulu has given way to cordial and friendly relations. Heartened by the change, Miertsching has again felt the urge to proselytize.

One evening, McClure discovered the brother in the act of distributing tracts on matters of religion and right conduct among the crew. Observing the drag such moralizing has on the spirits of his men, he ordered the ship's fiddler to play a tune and festivities to commence. The men danced the *hornpipe*, a traditional sailor's dance that requires neither women nor much in the way of room. It is performed solo, in place, with arms folded across the chest and lively, stomping feet. As if to taunt the brother, McClure made a point of rewarding one particularly gifted dancer and singer with a large glass of wine.

While there was no sign of the *Enterprise* on the voyage, the *Investigator* did encounter HMS *Plover*, under orders to remain in Kotzebue Sound as a western depot for Franklin and his men should they wander out of the wilderness. The ship had been in this position for two years, and a great number of natives had reportedly gathered on shore. Hoping the British seamen might have seized this opportunity to acquaint these people with the gospel, the brother learned instead that both officers and crew have shamelessly availed themselves of the hospitality on offer. So much so, he fears an Anglo-Esquimaux colony will soon result.

There, the warship HMS *Herald*, under the command of Captain Henry Kellett, was also met. Letters and supplies were exchanged, along with three muscular volunteers who joined the *Investigator*'s crew, short-handed since Honolulu. They were welcomed aboard with a final recruit, Mongo, a pup obtained from the resident Esquimaux. There was no word on the whereabouts of the *Enterprise*, but even a single day's delay so late in the season might well cost them the better part of a year. The crew readied the tools of ice navigation: ice anchors, chisels, hatchets, saws, and hawsers—heavy ropes or lines. The crow's nest was brought out and hoisted into position.

McClure continued to claim that the *Enterprise*—which he knows set sail from Honolulu on an indirect course around the Aleutian Chain—had surely passed this spot and was already further north. This raised eyebrows among his officers. McClure told Kellett that he had orders to engage the ice by August 1—there was not a moment to lose. Kellett found it difficult to believe that the *Enterprise* had passed this way without his notice. As ranking officer, he urged McClure to wait forty-eight hours for his consort. Through signal flags between ships, McClure declared that he could not afford to wait. On the bright night of July 31, the *Investigator* set sail north out of Kotzebue Sound to meet her fate alone.

From that moment, McClure has sensed that success is within reach. If he has indeed beaten Collinson's *Enterprise* to the Polar Sea, he likely has the best chance of solving the mystery of the Northwest Passage of any man on Earth. To the victor goes the fortune of a £10,000 parliamentary prize and the attendant fame and glory. McClure also has the slight and diminishing chance of playing saviour to the Franklin expedition and restoring Britain's Arctic hero to an anxious world.

Robert McClure is much less given to sentiment than his Esquimaux interpreter, the earnest Brother Johann Miertsching. Having never met his own father, who died five months before he was born, he is accustomed to the rigours of self-reliance. Under the guardianship of John Le Mesurier, his father's comrade-in-arms, he was educated at Eton College and Sandhurst Military Academy before entering the Royal Navy at the age of seventeen. On his first voyage to the Arctic, he served as a mate aboard HMS *Terror*, which became icebound in Foxe Channel, north of Hudson Bay, in 1836. Then, just two years ago, McClure served as first lieutenant aboard the *Enterprise* in an expedition to rescue Franklin led by Sir James Clark Ross. That time, Ross sailed aboard the *Investigator*. Both ships became caught in Lancaster Sound, where they overwintered in the ice. McClure is no stranger to polar navigation. He is also intimately familiar with the relative capabilities of the *Enterprise*—his current consort and rival. He counts on the fact that he has beaten his commander to the shores of the Polar Sea.

Morning mist and cloud recede, revealing the ship's location. Unbeknownst to McClure or his officers, they have rounded Point Barrow and now find themselves east of the cape, further along this northern shore of Russian America than any ship has dared to go. For the men of the *Investigator*, this realization brings a surge of pride.

The ice, however, continues to shift and squeeze. McClure is forced to dispatch five boats, with crews of eight men each, to tow their ship through the floes. Little progress is made. At night the men return to the *Investigator* so exhausted that even the fiddle is given a rest. McClure, wide awake under the midnight sun, seeks out his Esquimaux interpreter. As the crew sleeps, he wanders down to Miertsching's cabin for a remarkably frank and fluent discussion. He has grown fond of the brother's company.

McClure holds forth on what he believes to be the appropriate place of religion onboard a British ship. Simply put, no man at sea can be expected to maintain the attitudes and rigour of land-based Christianity, which calls for meekness, regret, and morbid self-reflection. It offers little in the way of mateyness. Surrounded by constant perils, removed from the moderating influence of society, a man must have courage—not hang his head in shame. The edge of the Polar Sea offers little release from the pressures of bare survival. It is a place for bold action, not pious introspection. Camaraderie, esprit de corps, will see them through.

This direct attack on the brother's efforts is not unexpected. The captain has more than once embarrassed him in front of the men with calls for less puritanical reading and more singing, dancing, and drinking. In response, the brother stiffens his back and reflexively quotes St. Paul. He concludes by saying, "I would gladly call myself and really *be* a Christian."

"You are not yet a true seamen or you would have other views," McClure replies. "Mr. Marx, who was formerly a lieutenant on a number of warships, had adopted your sort of land-Christianity. He learned by experience that it does not serve onboard ship. So he gave up the sea, became a parson, and now writes tracts for old wives. You should have given the leaflets you've been distributing to lost women. They would have given you more thanks than my sailors."

Both men enjoy these debates, McClure for his innate need to provoke, challenge, and win. He can be himself with this foreigner, this man who lives outside the naval tribe, unconnected to the chain of command that winds its way through the soul of every man onboard this ship and those under sail around the globe. Miertsching is a civilian who, by all rights, should not even be onboard his ship. He is a man with whom McClure has almost nothing in common and of whom he has few expectations. He is becoming a kind of friend.

For his part, the brother prefers this kind of encounter to polite toleration. Better to openly debate the merits of faith than accept his tracts with a civil nod and then stuff them in the stove. This is a dialectic he is well trained for. He is equipped with a series of proofs upon which he has grown to rely. As he finds himself again roused to quote scripture and other authoritative texts, McClure abruptly ends the discussion by handing the brother his guitar.

"Sing something," the captain orders.

Miertsching opens his hymnbook and continues the debate musically, with a tune titled, "How great a bliss to be a sheep of Jesus, and to be guided by His shepherd staff."

The captain listens to three verses before rising to thank his host and wish him a good night.

At three o'clock in the morning, the brother is awoken, told to dress, and ordered to immediately report for duty. His services are required on shore.

Three unarmed men stand in the skewed Arctic light. They are confronted by a foreign craft bearing pale, ghostly figures clad in what appear to be the skins of unknown beasts. Behind them is an even more unsettling sight: a crowd of similar beings crawling over what can only be described as some kind of living island. The men remain transfixed by this vision until at last one of them finds the courage to call out a greeting.

For Miertsching, one of the intruders seated at the bow of the approaching boat, the sights and sounds on shore momentarily transport him back to the distant coast of Labrador and its people, whom he grew to know and love. The men before him now possess the same features, manners, fur clothing, and sealskin boots as the Esquimaux at the other end of the continent. Although they speak a different dialect, he easily understands their language. Miraculously, they understand him. This is cause to rejoice. It means there must be some communication between these Arctic peoples—and that must mean a passage.

Europeans began using the word "Esquimaux" to identify aboriginal people encountered in Arctic and sub-Arctic Greenland, North America, and Siberia in the sixteenth century. Originally an Algonquian word, it was long believed to mean "eaters of raw flesh," although it probably made reference to snowshoes. The self-designation of these peoples varies across regions and dialects, and includes Inuit, Inupiat, Inuvialuit, and Alutiit, each of which means "the people" or "the real people." As a group, they are linguistically, culturally, and biologically distinguishable from North America's other aboriginal peoples. Their lives are completely adapted to one of the world's most extreme environments, and their presence on this continent dates back at least 3,800 years. They are viewed by Miertsching's companions with a mixture of awe, suspicion, and pity.

The brother, the ship's surgeon, and seven seamen step ashore. They are greeted affectionately, each Esquimaux rubbing noses with each startled European in turn. The Esquimaux lead them to a small hillock where tents can be seen in the distance. They urge the brother and his companions to accompany them to meet their families. When the newcomers hesitate, the Esquimaux inexplicably flee—leaving the Investigators on a vast plain covered in low grass, lichens, and small Arctic flowers. There is no tree or bush, save the Arctic willow, which attains a height of only seven inches in the most ideal conditions. The sparse vegetation reaches from the cool, shallow soil toward the temporarily generous sun.

Although this landscape lets the men momentarily forget their battle with the ice, in truth it is never far away. Permafrost, perennially frozen ground, lies mere inches beneath the soles of their boots. Widespread across the northern part of the Northern Hemisphere, and likely all of Antarctica, permafrost probably first occurred at the onset of glacial conditions about three million years ago. It covers approximately a quarter of the planet's land surface. In the far north, permafrost is present nearly everywhere except under the lakes and rivers that do not freeze to the bottom. Here on the north shore of North America, it can reach thicknesses of twenty-five hundred feet—and twice that in northern Siberia. It locks away vast, untold quantities of methane gas, preventing it from being released into the atmosphere. Remarkably, a kind of permafrost even occurs *under* the Polar Sea. While the existence of permafrost is well known to northern peoples, Western science did not begin its study until 1836, when Russians working in Siberia took an interest. Of permafrost, British scientists in 1850 know next to nothing.

Surveying the scene, Ship's Surgeon Alexander Armstrong notes its suitability as pasturage. A man of science and reason, he is also the ship's naturalist. Born in Donegal, Ireland, Armstrong attended Dublin's Trinity College and then the University of Edinburgh. He became a doctor of medicine at the age of twenty-three and joined the navy the following year. He served in the Mediterranean, North and South America, the West Indies, South Pacific, Africa, and Asia before being elevated to Fleet Surgeon last year. Now, at age thirty-two, he finds himself on a bold new adventure near the top of the world.

Tall, handsome, and athletic, Armstrong strides across the land. His expression is confident and determined; his physical form imposing. While this combination gives most men pause—and should afford him quiet confidence—he is instead willful, superior, and highly combative. Despite his natural gifts, his academic and professional achievements, Armstrong behaves as a man with everything to prove.

He barely considers the existence of the Esquimaux interpreter. Miertsching, for his part, harbours no warm feelings for the surgeon. While the brother is pleased that the bitter struggles that flared early in their voyage have largely cooled, he perceives that—beneath the ashes—the embers continue to glow. He knows that the captain has little confidence in his officers, who are quick to criticize his decisions. The brother, who wants nothing more than peace, feels caught between them and McClure. Although most officers are friendly and treat him with respect, Armstrong is cold and aloof. The brother remains careful in his presence, when his company cannot be avoided altogether.

The Esquimaux do not appear to be returning. The Investigators gather stones to build a cairn, and then scratch through the thin, thawed surface. In a bottle, they enclose news of their heroic search and bury it against the frozen earth.

Back at the *Investigator*, the ship's company is met with a party of fifteen natives in a long skin boat paddled by women. The brother recognizes this as an *umiak* or "women's boat," as opposed the smaller covered *kayaks*, which are piloted by men. These people are invited up onto the deck of the *Investigator*, where the crew partakes in the fascinating and highly amusing custom of nose-rubbing. The guests begin to dance and sing.

Although familiar with the use of firearms, the Esquimaux are armed only with large metal knives, spears, and arrows. They explain that they trade with foreigners (Russians) and that two previous ships have appeared in the west. None, however, has reached as far as the *Investigator*. Miertsching distributes gifts of tobacco, scissors, knives, and beads. He in turn receives skins and mitts. To him, these people appear ragged and desperately poor. The women's faces are tattooed with vertical, parallel lines extending from the lower lip down the chin. The men wear their black hair cut short at the crown, and large glass beads or round stones inserted into holes cut in their lower lips. First worn by boys as they reach adolescence, these ornaments, or labrets, thoroughly disgust Armstrong. He cannot abide rubbing noses with a people he finds filthy beyond compare.

Soon, two more umiaks arrive, carrying Esquimaux with fish and seabirds for trade. Armstrong takes the opportunity to measure these folk and record their dimensions, a process they find highly amusing. One man sells the same caribou skin to three unwitting sailors. When McClure's back is turned, he feels a hand slip into his pocket in search of treasure. The Esquimaux have been well paid for their fish and birds, and given gifts besides. Regardless, the thieving spreads and can only be stopped by physical force. The Esquimaux manage to make off with numerous items, the loss of which is noticed only after they are ordered off the ship and their umiaks have disappeared among the floes.

Three days later, the *Investigator* sails close to a low, gravel island. Esquimaux are again spotted on shore. Miertsching is summoned to accompany the captain, the surgeon, and two other officers, landing under the dome of clear blue sky and the snap of an English flag.

This time, the Esquimaux stand their ground with spears drawn, bows cocked. They launch a few arrows over the heads of the intruders. These men will fight to protect themselves—as will the well-armed Investigators. The safety of everyone is suddenly in the brother's hands. As the need for peace calls out, he instinctively places himself at its service. When the Esquimaux hear his words, a version of their own language, they set aside their fear and welcome the strangers, rubbing noses two or three times with each man in turn.

The Esquimaux are camped in a cluster of tents on a knoll above the beach. Their leader, Attua, is a friendly old man with three wives and thirteen children. In comparison to the last Esquimaux met, to Miertsching they appear clean and well clothed. Have they any news of Franklin? They know nothing of him or any other European. The brother and his companions are the first *kablunaks* (white men) they have seen. As they speak, they are unable to take their eyes off the *Investigator*. Although she is a full half-hour's paddle away, each shift and movement of the ship elicits a fresh jolt of alarm. They ask the brother if he has arrived on a kind of swimming monster.

In spite of their evident innocence and fascination, they are not untouched by Western civilization. Among their possessions is a musket, carefully wrapped in skins, with the words BARNETT LONDON, 1840, engraved on the lock. The weapon was obtained through trade with natives to the south. However, they have neither lead nor powder.

The Esquimaux report that open water occurs along this coast for only two moons each year. There is little time to lose. Even as the brother gathers information that could be vital for their survival, he seeks an opportunity to broach the subject of religion. He learns that, although they have no concept of a god that can be reconciled with his own, they do believe in two supernatural beings: one good, the other evil. These beings dwell in separate regions where people go after death in accordance with how they conducted their lives here on the shores of the Polar Sea.

For the brother, it is a start. Convinced beyond the possibility of doubt that he knows the full and essential story, it pains him not to have time to enlighten and correct. Surely, having made this first, brief contact, other brethren will follow and these Children of Nature will soon come to know that their sins have been forgiven.

3.

THE GREAT BLUE CHEST

The day struggles to stay above freezing. Under heavy fog, a contrary wind sets the floes and cakes in motion, smashing into the hull until the ship is stopped, surrounded, and the ice anchors are set. The officers kill time by shooting into the mist. In a sudden wave of generosity, McClure gives Miertsching a double-barrelled rifle so he can join in the fun. However, the gift does little to assuage the brother's melancholy. He retreats to his cabin and imagines himself back home—not in the arms of a Saxon sweetheart, but in the company of the faithful.

All diversions and daydreams come to an end when the keel plows into a sandbar and the tide drives the *Investigator* over on her side. Unsecured objects roll and crash against bulkheads and decks as the men scramble to keep their balance. McClure gives the order to lower the boats and lighten the ship by transferring the heavy stores. When sail is set to regain control, one of the boats suddenly capsizes, sending eleven barrels of precious salt beef tumbling to rest on the sea floor five fathoms below. A stock of 3,344 pounds of protein, nearly 6,000 rations—enough to feed sixty-six men for over three months—has slipped beyond reach. Should the ship become trapped in the ice, this loss could come to haunt them. Witnessing the catastrophe, Ship's Surgeon Alexander Armstrong—the man responsible for the health and well-being of the crew—is convinced it could have been avoided. One of his fellow officers had sensibly suggested towing the heavy boats to the leeward rather than the windward side of the ship, but this advice was stubbornly ignored.

It is late in the evening before the *Investigator* is again afloat. When the sun at last appears through the mist, the view from the crow's nest shows no open water but unending miles of heavy ice. And then the rain sets in.

In the dense pack, progress slows to a halt. Some of the floes are thirty feet thick, with pressure ridges—areas where floes have been upended or forced atop one another in violent collisions—now frozen in great, solid masses. Composed of a "sail" of sheet ice that can soar up to forty feet above the waterline and a "keel" that can reach eighty feet below, they are a testament to the chaotic force of the Polar Sea, and a warning to any vessel that would venture here.

The men anchor their ship near a small group of low, sandy islands where they find two whale skulls and fresh polar bear tracks. While there is little in the way of vegetation, and no tree worthy of the name growing within hundreds of miles, there is plenty of driftwood. These large logs find their way down the various rivers and into the Polar Sea. An important resource for the people who live here, it is a gift from an alien land. The crew piles it up into a great bonfire. With his new rifle, Miertsching joins the officers in shooting at the numerous birds, including eiders—a species he recognizes from his years in Labrador.

The brother sets his gun aside to gather billowing wads of feathers from empty eider nests. In his hands he cups the soft, rich down, which is justly renowned for its ability to retain warmth. Common eiders are large sea ducks that range across the northern coasts of North America, eastern Siberia, and northern Europe—including the seacoast closest to Miertsching's childhood home. Like the walrus, common eiders prefer to dine on molluscs and clams. Plain, dun-coloured females are unadorned, but male eiders are strikingly beautiful, with pure black and white plumage, a bandit-like mask around their eyes, and a dashing jade-green nape. Like most birds found at these latitudes, they have come to avail themselves of the temporarily accessible feast, to court with balletic displays of poise and grace, and to breed—in the case of eiders, in colonies of up to fifteen thousand individuals.

Some of the Investigators know these birds by another name. Off the coast of northeast England, an ancient colony of eiders endures on the Farne Islands. There, they were granted protection from overzealous hunters and down gatherers by what may be the earliest bird conservation law, established in 676 by Saint Cuthbert, patron of Northumbria. The birds were chosen as the county's emblem and became known as Cuddy's ducks.

The air, dense and humid, grows increasingly cold. It is time to go. In addition to driftwood and eiderdown, the men load an empty net back into their boat. After ten hours in the attempt, not a single fish was caught.

On their way back to the *Investigator,* the party rows through a thin film of grease ice spreading like a stain upon the sea. For most of the men, this is the first time they have seen salt water freeze. It is a sure sign of summer's demise.

Five full days pass before the ship is free and underway in the falling snow. The unnerving shock of ice striking the hull increases in frequency. Sometimes it results in a screaming, metallic scrape; other times a loud crash followed by tumbling blows. Still, a narrow lane, or lead, appears to open in the ice heading north and a strong wind urges the ship toward the heart of the Polar Sea. McClure declares that this is the moment to gather their courage and sail straight to the Pole.

The crew rises to the challenge and presses the ship through increasingly narrow leads until the copper sheeting is peeled back from the hull, leaving dark, oak-splinter bruises on pristine slabs of ice. The slightest widening of the lead brings hope, which is forced under almost as soon as it appears. For the men on deck, anxiety, joy, disappointment, and terror follow and repeat in turn. Below, the collisions dislodge and break heavy and well-stowed items. Twice thrown from bed, the brother endures a sleepless night—the thought of that final blow and swift onrush of water never far from his mind.

Morning brings more snow, fog, and rain. By eleven o'clock, the ship is jammed between immovable floes—some large enough to hold a city. For Armstrong, the ice begins to take on an architectural character. He sees ice temples, columns, and facades almost purposefully designed. It is as if Nature is intent on either astounding or taunting him with her power.

Hours of hard labour are required to free the vessel. Only then does Captain McClure finally face the obvious impossibility of their situation and abandon his plan of sailing due north. But it is too late; the ship is soon caught again. The mood of the officers and crew grows solemn. No one—including the brother—dares speak to him.

At three o'clock in the morning, the *Investigator* is finally freed by the sweat of crews labouring in five boats. They strain at their oars through falling snow until the ship finds a navigable lead. But there is no wind to fill the sail, no bird or seal to break the stillness. The only sound is the cool, rippling current against the floes and the warm, measured stroke of muscle propelling wood. The men are ordered to press on until they can row no more. After eight hours, McClure orders the ship anchored to the ice. After dinner, there is no fiddle, song, or dance. Save for the snoring of exhausted men and the occasional knock of ice against the hull, the ship passes a quiet night.

The following day, progress is made toward the mouth of the Mackenzie River before the ship is anchored to a floe twelve miles long. In his cabin, in the company of his Scottish servant and Assistant Surgeon Henry Piers, the brother quietly marks his thirty-fourth birthday. He has now lived longer than his saviour.

Much of the mainland coast has been previously explored by Europeans, including Alexander Mackenzie, who, in his search for the Northwest Passage, travelled the river he discovered all the way to the shore of the Polar Sea in 1789. He named it Disappointment River, as it did not yield the passage. It was later renamed in his honour.

On John Franklin's first overland Arctic expedition (1819–22), he travelled in birchbark canoes down the Coppermine River, over 400 miles to the east. There, he managed to map 550 miles of the North American coast but lost eleven of his twenty men to starvation, murder, and—it was rumoured—cannibalism. What is certain is that he and his surviving men were forced to eat lichen and their own leather boots, gaining him the sobriquet "the man who ate his shoes."

Franklin's second and more successful overland Arctic expedition was a trip down the Mackenzie River in somewhat sturdier boats to the shore of the Polar Sea. His was only the third European expedition to reach it. He turned back to winter at Great Bear Lake, but returned to the delta the following year and explored 610 new miles west on along the coast. One of his officers, Dr. John Richardson, travelled east from the Mackenzie along the coast to the Coppermine. Although Franklin never reached his goal of Kotzebue Sound, he gained a knighthood for his efforts. He arrived at the Mackenzie delta on August 16, 1825—almost exactly twenty-five years before the *Investigator*'s arrival.

Captain Robert McClure and his crew travel in a vessel of an altogether different magnitude. They are now farther along the coast than any ship has ever sailed, and they have only just begun. The whereabouts of Sir John Franklin, now on his third Arctic expedition, is anyone's guess. Across the north, the search presses on.

On the morning of August 24, a seaman focuses a telescope on the distant shore. A few miles away, he spots a group of people—one of whom he believes is wearing European garb. This news propels McClure into action. Could it be a member of the lost expedition? At the very least, whoever it is might be able to pass along a letter through trading networks that would eventually find its way into civilized hands. With his interpreter, surgeon, and six seamen, the captain sets out to meet them.

Under clear skies, the land now rises and rolls in hills and cliffs. As the Investigators approach the shallow shore, the keel bites sand and surf breaks over the bow. The men are forced to step out, hoist their heavy boat up and onto their shoulders, and carry it to the beach.

Meanwhile, the Esquimaux prepare for defence. They point their spears, draw their bows, and shout at the intruders to go away. Picking up a few of the arrows that land near him, Miertsching advances—all the while repeating words of friendship and peace. When this does not produce the desired effect, he draws a pistol (loaded with a blank) and fires into the air. This persuades the Esquimaux to lay down their weapons. After a few moments of the brother's soothing speech, they treat the Europeans as friends and invite them to visit their shelter, a circular tent composed of driftwood and skins.

The group's elder declares that the Investigators are the first Europeans they have met. They have no contact with the Hudson's Bay Company traders on the Mackenzie River because they have heard that they possess a kind of "firewater" that kills. They trade instead with Attua's people to the west. The man's son hobbles toward the group on a pair of sticks, having recently broken his leg while hunting. Upon examining the boy, Armstrong discovers that the leg has become gangrenous. There is nothing he can do short of amputation, a remedy the family declines. The harsh realities of survival in this environment leave little room for the crippled and infirm. The margin of error is thin. If the brother's experience in Labrador is any guide, he believes that when the family sees no chance of a speedy recovery, the boy will be taken to a lonely place where the dogs are unlikely to roam. There, he will be left with a little food and abandoned.

The surgeon notices a button dangling from the man's ear. Upon closer inspection, the word LONDON is found stamped on its surface. The man says he received it from strangers who arrived in the area, built a shelter and soon died. This sends the Investigators to a nearby point of land to determine if it could have been Franklin and his men. There, they find only scattered driftwood, a broken stone blubber lamp, and the skeleton of a kayak amid the purple lupine—the ruins of an Esquimaux encampment.

A few days on, while searching another stretch of shore, Miertsching and Piers find abandoned Esquimaux caches—earthen pits filled with the flesh of caribou, seal, and polar bear, and the blubber of beluga whale. Another pit contains a smaller cache of birds. Covered with moss and sand, this meat is kept close to the permafrost to preserve it for the coming winter. They find no trace of Franklin or any other European presence. For dinner, they shoot and roast sandpipers on a driftwood fire.

At Cape Bathurst, more news is found. A pair of Esquimaux women report that Europeans recently arrived and remained for just two days, leaving gifts of beads and rings. McClure deduces that these must be Sir John

Richardson, Franklin's old comrade and friend, and Dr. John Rae on their overland search via the Mackenzie. The women explain that they are alone today because their men are out hunting whales. The following morning, a party of Investigators sets off to find them.

Where yesterday the land was coloured with tenacious grass and flowers, all now lies under a seven-inch mantle of snow. The men row their boat through water so shallow they repeatedly touch bottom. Fifteen miles into the journey, McClure declares that no proper seal, whale, or Esquimaux would be caught in water less than waist deep. The men are forced to get out, turn the boat, and begin dragging it back the way they came. The brother takes the opportunity to pull out the telescope and survey the horizon, sighting what appear to be man-made shapes and movement. On closer inspection, thirty tents, thirteen umiaks, and a multitude of kayaks are seen.

The response from the Esquimaux is swift. They stream toward the intruders with knives, harpoons, and bows drawn and ready. The men are followed by women armed in a similar fashion. With a chilling cry, they fire a volley of arrows. Twice McClure cries out, "What is to be done?"

The brother hands his rifle to the captain and pulls on his "Esquimaux frock," the caribou skin jacket he has taken to wearing on such occasions. He runs directly toward the advancing Esquimaux. Along the way, he draws a pistol and fires into the air, shouting to them to throw their weapons down. They continue their advance. The brother firmly plants his feet in the snow and holds the now-empty pistol in his hand.

"We are friends!" he proclaims. "We bring gifts. We mean you no harm!"

Hearing their own tongue, seeing a man clad in a somewhat familiar manner, they suddenly calm. McClure now steps forward and stands with his interpreter. In a dramatic gesture, the brother draws a line in the snow. After a soothing stream of assurances, the Esquimaux finally lay their weapons down. Anger, fear, and distrust disappear as they bring their women and children forward to view the strangers. Soon, Esquimaux infants are laid in European arms.

Armstrong notes that these compact, healthy-looking people wear clothes made of the finest skins. He admires the muscularity of the men and the beauty of the women. He is happy to see that the men do not wear labrets, although a few instead wear a piece of ivory through the septum of the nose. He tallies the features and charms of one young woman who clearly excites his interest: *good complexion, large, dark, sparkling eyes, beautiful pearl-like teeth, aquiline nose, a most luxuriant crop of raven-black hair, small and delicately formed hands and feet . . . pleasing features radiant with smiles of cheerful good-humour.* When he moves to offer her a present, she shows him her two-day-old baby. He turns his attention elsewhere.

Miertsching observes that, strangely, some of these people have brown hair and blue eyes. They refer to themselves as "the real people"—*Inuit.* All other folk to the south, some of whom they meet in trade, they call Indians. The Indians in turn trade at Fort Good Hope on the Mackenzie River. They know nothing of the Franklin expedition. Hoping to get word of his progress back to London, McClure showers gifts on their leader—a grey-haired old man named Kenalualik—to be considered as postage for forwarding a letter to the Hudson's Bay station. The gifts include a gun, one hundred rounds, and a demonstration of their use.

While the captain sets about surveying the area's flat topography, Miertsching takes the opportunity to learn more from his hosts. Where words fail him, the Esquimaux are ready with gestures and signs. The men have recently returned from a successful beluga whale hunt, perhaps the most important event of the year.

For these people, the small, toothed whales are a critical source of sustenance. Gregarious and highly vocal, beluga whales are known among English mariners as "sea canaries" for their remarkable repertoire of high-pitched clicks and whistles. Beluga whales can reach fifteen feet in length and are often rotund. With a smooth, bulbous forehead and what appears to be a smile, they are reminiscent of enormous and well-fed infants. *Beluga* is the Russian word for *white*, which aptly describes their bright, milky complexion. Most live their lives among pack ice in Arctic and sub-Arctic coastal waters, migrating south along the edge as the sea freezes over completely. Others remain in the pack year round, finding leads and polynyas (patches of open water) in which to breathe. They travel in groups of five to more than a thousand and prey upon a variety of fish, crustaceans, and worms. They are in turn sought for their nutritious flesh and rich, highly caloric blubber.

Men hunt the whales from kayaks constructed of sealskin, sewn with a bone needle and caribou tendon, over a whalebone frame sixteen feet in length. Inspecting examples of these watercraft, Armstrong finds they weigh only about forty-five pounds and can be carried over his shoulder with ease. From the seat of their kayaks, men throw harpoons tipped with bone, flint, or iron. After piercing the whale's smooth back, the barb detaches from the shaft. Attached to the barb is a sealskin bladder, filled with air, that hampers the whale's descent. They then follow their quarry, repeating the process, until it finally succumbs. Proof of a man's hunting prowess is sewn into his clothing. These badges include the shiny heads of loons and—as among European royalty—the pure white and black-tipped furs of ermines. Permanent marks can be read on a man's face in lines tattooed from the eye down across the cheek—one line for each whale taken.

The Esquimaux are curious to know if the Europeans are married. Here, the brother advises his companions to answer in the affirmative. He deems this falsehood necessary as these people are highly suspicious of any unmarried man, a state only appropriate for boys. For them, a man's greatness can be judged by the quality and number of his wives. The young man who was successful in this most recent hunt has just received his first tattoo—along with a second wife.

The brother now takes a moment to inform these people of an all-knowing spirit that dwells above the sun, moon, and stars, looking down upon his creation. The audience stares on in rapt amazement. They sit down and listen to all he has to say before Kenalualik offers polite correction.

"Above us is a great blue chest," the old man explains. "It is the sun's house. In daytime and throughout the summer, the sun is not in his house. When he goes there, the land grows dark. But when he is there, at night and in the winter, he can view the water and land below through clean-cut little holes. These are the stars."

They tell the brother of an afterlife similar to the one he has heard before: two spirits dwelling in separate lands, one good and the other bad. The good spirit looks after the animals, the other does harm to men. If, in this life, an Esquimaux feeds and clothes widows and orphans, he goes to that good land where the sun always shines, a place where there is no ice and it never rains. There, the animals are tame and easily caught by hand.

McClure returns from his survey and calls his men to the boat. As Miertsching rises to go, Kenalualik holds him back. "Wait and tell us more," the old man says. "Live with us."

The brother explains that he must go back to the ship, that they are seeking friends lost in the ice.

Kenalualik offers him a sledge and a team of dogs, seemingly the same breed as in Labrador. "With these," he says, "you can go to the ship across the ice when the next moon disappears and the sea is frozen over." When the brother declines once again, the old man repeats his generous offer—with the addition of a tent.

"My captain is calling. I must follow and obey."

And then Kenalualik summons a beautiful sixteen-year-old girl, his daughter, whom he sets before his guest.

"Take her," the old man says.

McClure grabs Miertsching and Kenalualik by the arm and leads them both to the boat, where gifts are being distributed. The brother lavishes extra presents on the old man and his daughter, taking care to ensure she receives sewing needles, which are highly prized.

Meanwhile, a woman is caught stealing the boat's compass, which she attempts to hide in her bosom. Then the brother is relieved of his handkerchief, which he quickly recovers. As the Investigators push away from shore, the crowd presses in—grabbing at the gunwales, trying to climb aboard. The strangers are then followed with an enthusiastic escort of seventeen kayaks and umiaks that see them all the way back to their ship.

Fifteen Esquimaux appear the following day and are invited aboard the *Investigator*. They range over the vessel, inspecting everything in their path, even following the brother to his cabin. They are anxious to see the crew's women and want to know where they are kept. McClure, disregarding their known fear of firewater, invites a man to his cabin for a glass of wine, which is tasted but not consumed. The Europeans proceed to test their guests with every form of drink onboard, but they will take only water. In a similar experiment, various foods are supplied but all are rejected with the exception of salted pork.

The crew delight in trading clothes with their guests, pitying them in the exchange. The Royal Navy has yet to establish a uniform for seamen, leaving them responsible for their own clothing. Crewmembers have either brought their own clothes, made them from cloth provided onboard, or have purchased slops—cheap, ready-made clothes—from the purser. They wear cotton underclothes, wool trousers, jackets, and overcoats. Their boots are made of leather. The Esquimaux eagerly trade away fine, caribou-skin parkas—worn with the fur facing inward—and caribou-skin trousers. Their hooded parkas are designed without a front opening, to minimize heat loss. A superior insulator, caribou fur is ideal protection for the harsh Arctic climate. In winter, two layers are worn for added insulation. They wear sealskin boots, for durability and water resistance. Inuit clothing—far superior to that worn by the visiting Europeans—is one of the keys to their survival. Children wear miniature versions of adult clothes, while infants are kept in the loose hood of their mother's parka, close to her warm skin. In a demonstration, a previously unseen baby is pulled out by its feet and proudly displayed.

This causes another woman to weep. She explains that while she was out collecting mussels and sea grass along the shore, she left her own baby to play with pebbles nearby. She heard a scream and ran back to find a polar bear dragging it away in its jaws. The bear swam to a floe and proceeded to rip apart and consume the child as she looked on in horror.

While McClure suspects his guests may be inventing stories in the hopes of some compensation or reward, the brother is not so sure. He tries to console the woman with his description of a great spirit. He tells her how she should live, and that if she follows his words she will be reunited with her child in the next life, in a place where there are no sorrow or tears.

Where the Investigators are headed next is as far from that warm and gentle place as the brother can imagine. Even these Inuit—the last people at the edge of the world—tell him they don't know what lies to the north beyond the ice. They are certain of only this: it is the Land of the White Bear.

Before departing the inhabited cape for thicker ice, for deeper Polar Sea, the Investigators encounter a disturbing phenomenon. Thick columns of smoke are seen rising from a cliff eight miles down the coast. Remarkably, a seaman reports that he has seen tents and men with white jackets near its source. Could this be Franklin's signal fire? A boat is lowered with Miertsching, Armstrong, and eight other men. Shortly after they paddle away, the brother looks back over his shoulder to see a terrifying sight: smoke billowing from the *Investigator.*

How fully, completely he now feels love for this ship. She has come to represent life itself. He and his companions in the boat don't know what to do. If the ship burns or the gunpowder sparks and explodes, they will be stranded without provisions and will most certainly perish. They watch in silence as the smoke onboard slowly diminishes and then disappears. At last, they turn their attention to the smoking cliffs before them.

After a two-hour row, they find no sign of tents or men, only thick pillars of smoke and sulphur stench wafting from vents in the earth. It is impossible to approach the plumes closer than ten feet without burning their boots. Where at Cape Bathurst the land rose only sixty feet, here the cliffs reach up to four hundred feet above the tide. The land itself appears as a mass of pumice-like rubble in various shades of singed ochre. When the men poke their oars into the ground, the fumes increase although no flame can be seen.

Where the brother might be reminded of a biblical description of the netherworld—a place of eternal fire—he sees instead a vast, natural laboratory, a place where things are rendered down and remade. Along the shore, vast deposits of brown coal (lignite) smoulder as the cliffs erode and fresh minerals are exposed to the air. These hills have been burning for many centuries, as they will for centuries to come.

The men return to the ship with samples of the substance, carried in scorched handkerchiefs. When laid atop the captain's fine mahogany table, they burn marks into the surface.

McClure informs them that the fire they saw onboard earlier that day was caused by a problem with the ship's primary heating apparatus, her Sylvester stove. Although it was quickly extinguished, some rigging was destroyed. There is no cause for worry. They have endured worse—and anticipate worse to come.

4.

POSSESSION

Night comes early now. Beyond the reach of star and moonlight—cradled by a small forest of wood—Johann Miertsching lies afloat in darkness. And yet the hard white intrudes whenever contact is made. A strong wind drives the *Investigator* into a collision so sharp the brother is thrown. He is not alone. His fellow officers find themselves similarly ejected from their berths. Shocked from sleep, ajolt with fresh adrenaline, each man finds himself with arms outstretched, reaching for steadying bulkheads, listening for the cries of distress and rush of water that fail to come. Hearing no such alarm, the brother crawls back into bed and begins the process of gathering his wits. Soon, ship and ice again collide, and he finds himself tossed once more. He picks himself up, wondering what possessed him to agree to such a voyage, questioning whether he had been listening to the will of God or the whisper of his own vanity all those months ago, guessing when it all might mercifully end.

In the bright morning sun of September 6, 1850, the brother defies his lack of sleep to rush up to the outer deck and reconfirm the previous day's discovery. And there, just two miles past the bow, the unknown land awaits. This was no dream. The cliffs that yesterday could only be seen through squinted eyes now rise sharply from the beach to eight hundred feet. Beyond, even loftier peaks soar to perhaps two thousand feet over bare, denuded land. By ten o'clock, the ship is at anchor and a beaming captain McClure musters his yawning officers. They launch a pair of boats and row north to take possession of this land for their Royal Mistress.

En route, they dodge cakes and floes until forced out of their boats to haul them across the ice. A final stretch of open water allows them to relaunch the boats and row ashore in style. At last, they step over the gunwales and onto pebbled shore. They march past the high-tide line, unfurl the English flag, and gather round. A flagstaff is planted in the soil of this, one of the last unknown places on Earth. They claim it for Queen Victoria with three hurrahs, and then a fourth for good measure. They name the land "Baring" in honour of the First Lord of the Admiralty, a patron of their expedition. A scroll listing the names of the officers of the *Investigator* is then buried in a cask beneath a fifteen-foot pole, alerting any future passers-by.

The men swell with the pride of accomplishment. Much has been endured to make this discovery—surely the first of many to come. Watching the shock of red flutter on this bright blue day is the kind of reward for which they have been waiting. They pause for a moment over a nearby stream to drink the fresh water of Baring Land, spiked with rum, in a toast to the continued health of Her Royal Britannic Majesty.

A quick survey is in order. McClure dispatches his surgeon to gather specimens.

Exercising his long stride, Alexander Armstrong pauses now and then to inspect the peaty soil, sand, stones, and small patches of saxifrage clinging to the otherwise brown and sterile land. It is a landscape far more barren than the last they encountered, a place seemingly at the edge of life itself. And yet he finds the prints of Arctic fox, the antlers of caribou, the skull of a polar bear. Then, movement—a couple of startled Arctic hares bound for cover. It remains only for someone to climb the highest point of land. Someone fit, observant, and of suitable rank. Armstrong nearly makes it to the top before being called back to departing boats.

The brother has no interest in christening barren tracts of rock and ice with the names of haughty Englishmen. He is thankful for this chance to walk the land at the periphery of God's creation. His last minutes abroad this day are spent gathering flowers, plants, and stones for his own natural history collection. On the stroll back to the boats, he sights and shoots a pair of snow geese, which are as bright and pure as their name.

McClure is aware that this is not the discovery of a New World, the Pole, or even the Passage. This land will never be a trove of wealth for himself or his nation, a colony from which mariners will fill their ships with valuable resources or, like the Sandwich Islands, a paradise in which to rest and refresh. Still, it is a fine, curving line he has added to the chart of the known world. It will go some distance in helping posterity remember his name. He makes his way down to the shore and boats, allowing himself one last, lingering look at the ensign coming to life in the breeze.

That night, buoyed by an irrepressible good mood, the captain entertains the officers in his cabin—exposing them to that charming, congenial side of himself he allows out on occasions of success. Perhaps by sharing his enthusiasm, his pride in what they've already accomplished, their own ambition will rise and—faced with their combined powers—the Passage might reveal itself to them.

To the west, the ice—twenty to thirty feet thick—is in constant battle with the shore. Seeing no hope of passing along that side of the land, McClure sets a course up the eastern coast. The *Investigator* travels through varying fog into what appears to be a bay, ten to fifteen miles wide, with Baring Land to the west, Wollonston's Land to the east, and a narrowing stretch of ice and water between. Following the midday meal, two small, forlorn-looking islands are spotted standing firm in the streaming floes. As yet unnamed, they are models in miniature of the surrounding wilderness. The current pulls the ship north amid a regatta of ice, which keeps men awake at night with insistent reminders of just how far they've come.

The shore is now almost completely covered in snow. The temperature continues to fall until it reaches a mere 15 degrees on September 13, a full week before the equinox and the official end of summer. The *Investigator* is now stopped and held tight in the crowding floes. McClure considers options. Miertsching, who has caught a bad cold, retreats to his cabin for several days. Exhausted, he is jerked awake now and then at the crack of musket fire as the officers amuse themselves on the outer deck, shooting at the treacherous ice and any unfortunate passing gulls.

And then the floes loosen their grip, allowing room for hope. McClure orders all hands on deck to set ice anchors and pull hawsers, manhandling the ship through ephemeral leads. After several hours, and only inches gained, even McClure cannot deny the futility of the task. The cracks close and are quickly scarred over with new ice. McClure's ambition for heading farther north this season flickers and then dies. He reins in his ambition and now turns his full attention to getting his ship and her sixty-six men to some safe harbour for winter.

But this thought comes too late. The *Investigator* is trapped. The cold, low cloud and falling snow weigh down the spirits of the men. The breath of anxious sailors instantly freeze on exposed eyelashes, beards, and mustaches. Despite the weather, Armstrong—concerned for their health and morale—orders the men off the ship and onto the ice for their daily exercise. They tackle one another in the new fallen snow, wrestling like boys, sliding down the hummocks on their backsides, momentarily forgetting the peril they and their ship are in—setting aside the thought that they are now like Franklin, lost to the world on the edge of the uncharted Polar Sea.

The wind rises and sets the ice in motion. In several places, open water can be seen. Muscle and toil are again deployed to little effect. A day later, thin cracks expand and a lead can be seen connecting to open water farther north. The men are able to swing their ship about, pull her through to a widening crack, and then set sail—only to be stopped after the gain of several yards. And then, through no effort of their own, the fissure widens into a lead and the way opens until the ship is sailing, dodging ice, and—it seems—experiencing the swell of open ocean. If this is the case, it means that they are not heading deeper into a bay, but travelling a strait between two separate pieces of land. This would mean their newly discovered Baring Land, to the west, is in fact an island—and that could mean a passage.

Given the options, Ice Master Stephen Court is of the opinion that the *Investigator* is in the safest position possible, anchored to a large and sheltering floe. To cast off now would put the ship in further jeopardy amid the smaller ice. This is an opinion with which Armstrong and other officers concur. McClure, however, disagrees. His instincts tell him that their best hope of escape is to turn and seek a safe anchorage to the south. He is coming to the conclusion that the recently declared Baring Land is in fact "Banks Land"—described but never reached by Sir William Edward Parry and his men. Given this revelation, he suspects that they are now tantalizingly close to the completion of the Northwest Passage.

Thirty summers before, a corner of that uncharted land was first glimpsed by Frederick William Beechey, a lieutenant aboard Parry's ship, HMS *Hecla*. His view was from the slopes of Cape Dundas on southwest Melville Island, an island discovered by Parry on his first search for the Northwest Passage. His ship's surgeon, Alexander Fisher, recorded the sighting on August 7, 1820:

> *. . . land was seen to-day, extending from S. to W.S.W., and supposed to be about fifty miles off. Whether this is the continent of America, or an island lying off it, is certainly a question that our present knowledge is inadequate to decide; I shall therefore not offer an opinion on the subject. From the same elevated situation that this land was seen, we had also a good view of the sea to the westward, or rather, I am sorry to say, of the ice; for, as far as we could see . . . was covered with ice. There were here and there, indeed, small pools, and lanes of open water, but no continuous opening . . . the ice here is much heavier than any that we saw before, and is at the same time quite of a different character; for, instead of its presenting an even surface, like the ice in Baffin's Bay, it is completely covered with hummocks . . .*

The hummocks observed were mounds of ice that form from old pressure ridges that build and then melt down in summer, a process not unlike the long-term erosion of steep young mountain peaks into gently rolling hills. The sight of such thick and variable ice offered slim hope for navigation during the short summer season. The following day, Parry himself went to have a look, recording these impressions:

> *Everything was so quiet at nine o'clock as to induce me to venture up the hill abreast of us, in order to have a view of the newly-discovered land to the southwest . . . This land, which extends beyond the 117th degree of west longitude, and is*

> *the most western yet discovered in the Polar Sea to the northward of the American Continent, was honoured with the name of BANKS'S LAND, out of respect to the late venerable and worthy president of the Royal Society.*

Although Parry was unaware, this was the last, ice-choked stretch that stood between him and ultimate success in navigating the Northwest Passage. His single winter here in the western Arctic was a mere inconvenience. He returned to England to claim a £5,000 parliamentary prize for himself and his crew. In the end, Parry's early and near success in the pursuit of the Northwest Passage showed him to be a bold and lucky genius of polar navigation.

Were this open water, the Investigators could be at Melville Island in Parry's Winter Harbour in a matter of hours. But it is not so. The sea is choked with impassable masses of hardening ice, just as Parry and his men encountered thirty years ago. McClure gives the order to cast off and turn south.

Progress is made through similar means: ice anchors, hawsers, brute strength. Despite their efforts, the Investigators continue to drift north in the ice stream. At times, the sheets of young ice crack in two, one half climbing over the top of the other with a loud, chirping squeak. Other times, enormous older floes—more influenced by the currents and tides—rush past slower-moving neighbours with remarkable velocity, flipping and casting aside lesser ice caught in their path. In a sure sign of their struggle's end, the ship's battered rudder is pulled to save it from destruction.

On Saturday, September 21, the temperature drops to 7 degrees. The crew anchors the ship again to a large mother floe with several hawsers and a chain. There is little to do but wait. To relieve stress and forget themselves, the idled men pour their energies into song, dance, and drink, exhausting themselves in the mix. Some appear in masquerade, others perform skits and comic routines. The resulting racket is too loud even for McClure, who at last orders them to bed in time for a few hours rest before the observance of Sunday service.

By Monday, that same small pair of islands are again spotted, proving the ship has now been driven fifteen miles south, back down the channel. The ice immediately surrounding the ship has frozen into a solid, contiguous mass. Onto this surface, Miertsching and Armstrong descend.

The air is frigid and, unlike the atmosphere inside the ship, utterly odour-free. Starting with smell and touch, it is as if even the senses are being pulled down and away, subsumed in the ice and cold. The brother and the surgeon have little to say to one another, each caught up in his own anxieties. They walk in silence around the ship, inspecting the dark hull, the dense floe, and then a shelter created by some of the officers nearby. By excavating one of the larger hummocks, they have created an ice room large enough to house a dozen men—and they have ambitious plans for expansion. This happy diversion is soon ruined by the sound of the whip striking a man's back. The gunroom steward has been held in chains for several days and is now being flayed for an unknown offence.

Considering their critical position, feeling the weight of responsibility press down, McClure gives the order to assemble on deck a full year's supply of provisions, to be ready at a moment's notice should the ice shift and crush the ship. The crew might then be able to escape across the floes, make it to land, survive the winter, and then row to the mainland and eventual rescue. That is the plan, the faint hope. Has Franklin already been forced to make just such a journey? Would Franklin and his men be forced to rescue the rescuers, giving aid to McClure? The shame is too deep to fathom.

The men labour throughout the day, shifting loads until the deck is crowded with stores: woollen clothes, canvas tents, skins, blankets, sledges, firearms, ammunition, medicine, bread, potatoes, and *pemmican*—dried, ground meat mixed with melted fat. Observing the preparations, Armstrong is momentarily distracted by a flock of ducks passing overhead against the ashen sky, heading south to warmth and safety. Soon after their passing, the sun sets on the clear edge of the horizon, throwing up brilliant hues against the underbelly of cloud, making it appear as a sheet of molten metal.

Morning brings a change of wind, a falling barometer, and rapid drift to the south. Soon, the ship passes those two small islands mid-channel, and into increasingly hazardous positions. The integrity of their anchor floe continues to hold. From the deck, the men watch as other floes crash into the shore of Baring or Banks Land, releasing loud and violent energy on impact. Should they continue in their present course, the Investigators will surely find their ship shattered, their supplies tipped and pitched into the teeth of ice. Together, the brother, the surgeon, and the captain each observe the impending catastrophe, each alone with his fear.

By 9 a.m., their sheltering floe is threatened. They watch in horror, thoughts and words held in, robbed of the illusion that there is anything anyone can do. The passing ice strains the hawsers and chains—and then the beams of the ship start to groan. A loud, sustained grinding noise is heard close by, and then black lightning cracks appear in the their anchor floe and extend from the side of the ship. The fissures widen, multiply, and then suddenly the ship is floating free, heading toward certain destruction on shore. An hour of unrelieved terror passes, and then the remains of their floe run aground on a sandbar, their drift arrested, their fate forestalled.

And still, large and small masses of ice churn and tumble past. Surely, their lines and luck will not hold for long. The unrelenting pressure against the hull causes the timbers and planks to buckle. Chunks of tightly packed ice shove past like a mob in revolution. As the day fades, the wind picks up and snow starts to fall, obscuring the land the ship once again approaches. The men see the dim, barren shore as the site of their inevitable ruin.

Throughout the evening, the waterborne avalanche continues. McClure calls all hands on the upper deck and divides the officers and seamen into two crews, each assigned to a boat, so that they may escape to land. The brother pitches in, helping divide supplies. As they ready to abandon ship, the *Investigator* is twice lifted up and thrown broadside, sending men and supplies rolling. Somehow, she rights herself each time. By midnight, the tumult eases and the skies clear, allowing the moon to illuminate their desperate position. They gaze with gratitude on that battered but intact floe to which they remain—miraculously—attached.

The men are ordered back below to select their warmest change of clothing. In the rush, the brother corners McClure for a few private words meant to encourage and embolden. He has come prepared. To Johann August Miertsching, missionary of the Moravian Brotherhood, the life beyond this life is as real as the ice that gnaws and grinds, or the wood that shields them still. From under his arm he retrieves his Moravian Manual of Devotion and reads aloud:

"Be ye therefore ready, for ye know neither the day nor the hour when the Son of Man cometh . . ."

McClure pauses momentarily, listens with solemnity, then turns to the pressing tasks at hand.

With winter clothes and snow boots bundled and ready, the brother closes his cabin door to prepare himself, to ask that this test might pass him by. Or, if that is not to be, to beg for mercy.

Two hours later, he finds himself roused and mustered with the others on the darkened deck, clutching his bundle and biscuits, pockets filled with gunpowder and shot, ready at any moment to jump into his appointed boat, to row with all his strength through ice to a land of snow. If they have the grace to make it, he reckons, it will be less than a week before everyone has perished.

Merciful Lord and Saviour, who desires not the death of a sinner, but rather that he should turn from his wickedness and live . . .

At three o'clock in the morning, still huddled with the others on deck, Miertsching convinces himself that he sees the hand of God guiding both ice and ship. And yet the opposing floes squeeze so hard that the captain's door will neither open nor close. Drunk with spent adrenaline, the brother joins the other men and lies down upon the deck with his bundle under his head.

These are anxious hours. Lord, graciously protect us. And if it be Thy will that our journey end here, make ready our hearts. We are all so slack and weary that we yearn for rest.

Some kinds of sleep are debilitating, pushing the weary even further from any sense of rest. Such are the few hours spent shivering on the frozen planks. Miertsching rises to his feet in the dark, stupefied, unaware of the passage of time. He will stand ready with the crew through the balance of the night, waiting. There comes a point at which even terror breaks down, the worst result having been imagined and accepted. It is then that the wait—not the end—becomes the hateful thing.

The wind pushes the ice and the *Investigator* toward that pair of islands passed, then passed again. Its rocky, vertical cliff looms 120 feet above the grinding ice. Here, there is no beach. To forestall impact, the men work to secure six ice anchors and hawsers to the large, adjacent floe. All dozing men are now roused to their feet to watch as the ship passes within fifty feet of the opposing cliff—and yet somehow escapes.

Now great masses of ice—three and four times the size of the ship—collide and are forced on top of one another like great heaps of slag. As the piles move by and toward the ship, they shift and then topple with a low, thunderous roar. At times the pressure is so great, the beams so sprung, that oakum falls from the seams. In the hold below, casks—packed so tight they cannot shift—now begin to bulge and crack. The *Investigator* is hit from multiple directions, shoved broadside one way and then the next. She is heaved up and out of the water and then, with the collapsing ice, the ship is plunged back into the sea.

For seventeen hours they stand, eyeing their doom. Should any opportunity to abandon the ship now appear—to row through the ice or run across to shore—the men would surely bolt. But no route of escape reveals itself; all must be endured. And yet even endurance has its limits and depths. A number of sailors now approach that unseen frontier beyond which they cannot go.

The ways in which men confront their fate are as diverse as how they live their lives, falling into a spectrum with variation and extremes. Some calm themselves with faith in authority and protocol, some with their own recourse, others with mystical powers unseen. Most continue to work for group and self-preservation well past any indication that their efforts hold sway. For some, the break is sudden and sharp.

Men now abandon their posts, flee belowdecks, and break into the hold, where the spirits are kept. They throw themselves on the mercy of rum. Better to feel the familiar warmth, to deaden the senses, to laugh at or defy the inevitable. Over the shouts and threats of their captain, they fill their bellies with liquor, stupefying their senses in the hope of escaping the agonies to come.

Consigning them to their choice and fate, McClure returns to the deck to stand with the brother, the surgeon—the remaining officers and crew—to meet their greatest test yet. The *Investigator* is again forced up and over on her side, but this time a towering heap of ice threatens to bury the ship, forcing her under and down.

Then movement grinds to a halt. The tumult fades away. The *Investigator* remains on her side as the men struggle to stand upright, to grab hold in anticipation of the next shift and plunge. But the quiet lingers on. Cast out of the water, unable to move, the vessel is a beast exhausted. She can hold herself up no more. The men search each other's faces for some hint as to how they should feel, marvelling at—disbelieving—the sudden change. But the brother instinctively believes. In the silence, he hears the promise of God's mercy and command.

Thus far thou shalt go, but no farther.

5.

ARTICLES OF FAITH

The cold cannot be heard; it communicates strength in silence. Robed in white, the *Investigator* lies helpless on her side, the hoarfrost on her yards and rigging a fine, crystal lace. The weather is bright and clear; there is no breath of wind. The temperature has dropped to 2 degrees and hardening sets in. The quiet invades and takes hold of the ship. Words are soft and few.

It is impossible to walk upright on decks steeply raked. After pumping out more than two feet of water from the ship, little else can be done, save keeping watch and taking stock: six hawsers snapped, six ice anchors lost, one life buoy missing, numerous casks cracked or burst, a quantity of rum disappeared, all hands safe and accounted for. The crew mostly stay in their hammocks, which luckily swing to the preferred position. They sleep off the effects of intoxication and utter exhaustion. The officers keep to their cabins, taking what rest they can on berths askew.

The following day, September 28, brings little change,

other than the fact that the great field of ice now drifts north with the tide. It continues to thicken and grow.

Everyone now follows orders, leans into his work without complaint. The crew succeeds in righting the ship with block and tackle and the judicious use of muscle. But the *Investigator* cannot be put on a proper, even keel. She is not a nimble ship afloat at sea but a 422-ton edifice teetering in the rubble. Inside, the ship is damp throughout, including clothes and bedding, magnifying the ill effects of the cold. Despite their remarkable survival, their successful righting of the ship, or the fact that this is Saturday night, no man feels moved to dance or sing. Everyone awaits some movement or sign. With morning comes the order that all men are to immediately cease work. The entire company will muster outside on deck.

Under the flawless sky, the captain confronts the gaze of sixty-five addled men. They wait in silence, shifting from one cold foot to the other, venting steam. Following the official inspection, McClure retrieves a few leaves of paper and holds them in gloved hands. He reads aloud from the Articles of War, scripture of that secular religion universally observed and practised by the world's most powerful navy. Despite the cold, the men uncover and bow their heads to receive their rebuke.

> *. . . all persons in or belonging to Her Majesty's ships or vessels of war, being guilty of profane oaths, cursings, execrations, drunkenness, uncleanness, or other scandalous actions, in derogation of God's honour and corruption of good manners . . . Every person in the fleet, who through cowardice, negligence, or disaffection, shall in time of action*

withdraw or keep back, or not come to the fight or engagement . . . whomsoever shall be faulty therein, and shall not faithfully perform their duty, and defend the ship and goods in their convoy . . . or running away cowardly, and submitting the ships in their convoy to peril and hazard . . . If any person in or belonging to the fleet shall make or endeavor to make any mutinous assembly upon any pretence whatsoever . . . or shall disobey any lawful command of any of his superior officers; every such person being convicted of any such offence, by the sentence of a court martial shall suffer death, or such other punishment . . .

He then sets the rules aside.

"In our most critical time, in our ultimate hour of need . . . your behaviour disgraced yourselves, your fellow men, your country, and the Almighty who sought to protect you. You . . . are a band of thieves, unworthy of the name Englishmen. I am ashamed to know that such a thievish rabble walks the deck of an English ship. Those I refer to will be punished in accordance with the law . . . Tell me: Is the danger we have met, are still in, and continue to face not plain to you?"

He waits out the silence.

"We must rely upon one another," McClure at last concludes. "Moreover, it must be clear to everyone that only Almighty Providence, through a manifest miracle, has saved us from certain death—*despite* your actions."

Huddled in the audience, Miertsching is impressed with the substance and effect of the captain's speech. He thinks it was not only just, but hit all the right notes. And then, a surprise. He looks first to his left, and then to his right, and sees that no one can refrain from tears—even those weathered veterans of the sea. It is as if all the emotion held back and suffered over the past week has at last found an opening, has come welling up in one terrific gush. Perhaps this is the English way. The men's tears are followed by spontaneous calls of fidelity and the promise to mend their ways. The release culminates in an unreserved and emotional cheer for their captain, Robert McClure.

Chilled but unburdened, the men now make their way below. As they do, the three sailors who led the desertion present themselves and submit to the lash.

The ice sheet drifts with the tide. For vast stretches surrounding the ship it remains as a contiguous piece. Regardless, the men remain in a state of high alert, ready at any moment for the floe to fail and the battle to be rejoined. There is a growing agreement that for now, however, the struggle appears at an end.

Outside, the temperature drops to minus-five. Inside the officers' cabins it is 15 degrees with a relatively balmy 28 degrees in the gunroom, which serves as the officer's mess. Unless in a boil on the stove, or kept close to the skin, all water-based liquid is frozen.

The crew carries out its work with diligence. Coarse language is infrequently heard. The officers and captain are on the best of terms. Throughout the ship, there is an abiding sense of gratitude. Each day, this rubble field appears more and more likely to be the site of their winter home.

While the record of Arctic navigation dates back to the end of the fifteenth century, the number of crews that can claim to have actually overwintered in polar ice are relatively few. In the search for the Northwest Passage, William Edward Parry, John Franklin, John Ross, and James Clark Ross have managed such a feat. Although their trial-and-error methods date back just over thirty years, there is much to be gained by studying their journals. McClure has not only consulted the literature, he himself has spent two winters beset in polar ice: one aboard the *Terror* (1836–37) and one aboard the *Enterprise* (1848–49). He is under no illusions.

On his order, the topmasts and yards are lowered, the rigging stowed away. Above the deck, sturdy canvas is stretched and secured to the sides of the ship in the manner of a tent. When the storms and snows arrive, it will serve as shelter for exercise, to keep the men in shape and to beat back the effects of depression that are sure to come. But for now the skies are clear and there's work to be done. Here, near the middle of the pack, the ice still occasionally cracks and shifts. To provide some protection should she tip and fall on her side, McClure gives the order to clear the near rubble to provide a smooth landing surface for the rolling hull.

For this engineering project they are well equipped. Prior to setting sail from England, Ship's Mate Robert Wynniatt was instructed in the use of gunpowder for the purpose of blasting ice. This is a new concept in polar navigation, and everyone seems eager to give it a go. The first experiments are conducted on a small scale with a large audience, providing both entertainment and confidence for bolder moves. The nearby hummocks are targeted and destroyed, their residual parts used to fill in the nearby cracks and holes.

It is clear to McClure that this arsenal will no doubt prove useful in future engagements.

Since the defeat of France at Waterloo in 1815 and the end of the Napoleonic Wars, British naval supremacy has gone unquestioned. But with the great enemy vanquished, 90 percent of naval officers found themselves suddenly unemployed, struggling to survive on half pay, restless for the next opportunity for advancement and glory. While most able seamen had been discharged (reducing their numbers from 140,000 to 19,000), six thousand officers remained—along with ships and materiel. Into this vacuum of purpose surged the quest for the Northwest Passage. It has taken on the character of something between national sport and sacred crusade. When Franklin and his ships went missing in 1845, it transformed into a full campaign, complete with overland and marine assaults, territorial conquests and casualties, and comrades in need of rescue. In this struggle, the polar ice plays the role of both battlefield and enemy.

The quality and duration of daylight rapidly declines. Before the sun departs altogether, it seems intent on leaving the most dramatic and lasting impressions. For several days in a row, in the still evening air, a sign appears in the heavens, drawing onlookers out on deck. As the sun sets in the cobalt sky, it throws up a halo around itself with two or three partial or full suns arrayed to either side and below. The apparition shows a spectrum of colours ranging from red, near the inner ring of the solar halo, to a pale bluish tail extending out near the refracted suns on the outer edge. When the trinity of suns finally sets, a similar effect is seen encircling the full moon.

The optical phenomena observed around the sun are *parhelia.* Also known as mock suns or sun dogs, they are formed by the refraction of sunlight passing through ice crystals suspended in the air. Those seen in relation to the moon are paraselenae, or moon dogs. Since before recorded time, people have marvelled at their appearance and wondered what the remarkable signs foretold. Parhelia were mentioned by both the ancient Egyptians and Greeks. During England's dynastic Wars of the Roses, the decisive Battle of Mortimer's Cross was fought on February 2, 1461. Before the battle, a dramatic parhelion appeared with three complete suns. These were thought to presage the victory of—and show divine favour to—the three leaders of the House of York. The event was later referenced in 1590 by none other than William Shakespeare in his play *Henry VI, Part III.* The phenomenon was perhaps first clearly described by Jakob Hutter, founder of the Hutterites, in 1533. The radical Christian leader declared the celestial apparition he and many others observed in Moravia to be a miracle. For early navigators of the Polar Sea, the appearance of parhelia was thought to foretell ominous events and has been regarded with awe and terror. The educated officers of the *Investigator* are careful not to let the natural occurence become fodder for the superstitions of the largely illiterate crew.

The frequent appearance of aurora borealis, or northern lights, is more welcomed and seems somehow less open to interpretation. The shimmering sheets of light cheer the nighttime sky and draw universal acclaim among the men. In their colourful displays, they appear to connect the heavens and the ice with a broad cascade of fire.

With the enduring sub-zero temperatures, the ice gives the appearance of a landscape solid and set. But every now and again the men are sent a reminder that they are still, in fact, on the edge of the Polar Sea. Without warning, a huge chunk of ice shifts and strikes the *Investigator* on her quarterdeck, dropping her down and swinging her stern around. Before anchor lines can be set, the ice rises under her keel again, lifting her back to the surface, where she rapidly freezes in place. Following this minor adjustment, a seismic shift begins.

A low rumble is heard far in the distance. Like rolling thunder, it advances through the pack, growing so loud on approach that the officers must shout into the captain's ear. The men on deck watch in awe as the far fields of ice rise and fall, as if pushed up by an immense beast closing in on the ship. All around, the new ice fails and the hummocks fall and that old, sick feeling returns. Then the wave hits, and they ride out the wake until rocking stops, the sound fades away, and ice bears down again.

6.

THE LONG VIEW

The field of engagement expands across the roof of the world—from Iceland to the Aleutians, Hudson Bay to Hokkaido. Before it has completed its advance next spring, between five and six million square miles of sea ice will have formed throughout the Arctic, covering the Barents, Beaufort, Chukchi, East Siberian, Greenland, Kara, White, and Laptev seas. Beyond the Arctic Circle, the Sea of Okhotsk and the Bering Sea will also have been subdued.

The advance of the cryosphere is a seasonal offensive following the rapid retreat of the sun. Derived from the Greek *kryos*, for "frost" or "icy cold," the cryosphere is that frozen portion of Earth's surface that includes sea ice, ice sheets, permafrost, glaciers, frozen lakes and rivers, and territory covered by a mantle of snow. By January, fully 23 percent of the Northern Hemisphere will be claimed by frozen water—that solid, crystalline form of the classical element necessary for both the generation and maintenance of life. Surrounding the *Investigator*, old ice is incorporated into new, reaching down as well as out. After a prolonged struggle, the *Investigator* is at last entrapped in the frozen world.

Her crew, however, are still free to stretch and roam.

The party has been away from the ship since eight o'clock in the morning. The short day has been clear and bright, but the temperature has remained south of zero. In single file, McClure, Armstrong, Miertsching, Second Lieutenant Samuel Cresswell, and four seamen march across the ice and snow. They scramble up and over huge hummocks, armed with muskets, boarding pikes, a pickaxe, shovel, and flagstaff.

Between old hummocks and upturned floes lie sheets of flat new ice, formed in the previous hours or days. Sprouting across the new ice are fields of frost flowers that arrest the group with their beauty. Spaced in regular intervals, the featherlike structures bloom when saline vapour escapes from forming ice. They can grow as large as an English rose. Armstrong squats down to observe their crystals fracture the sunlight in prismatic displays. To him, the overall effect is of a carpet studded with gems.

Taking the lead, Armstrong approaches the thin edge of the floe to test its ability to bear the weight of the group. The ice suddenly gives way and he falls, immersing his legs in water before he is hauled out by the men nearby. After scrambling to his feet, he stands dripping in the sub-zero air, his wool trousers seizing up by the minute. He assures everyone—himself included—that he is no worse for wear.

The surgeon, along with several other men, begin experiencing the symptoms of hypothermia, that condition in which the body is unable to replenish heat being lost to the atmosphere as core temperature falls below 98.6 degrees. Hypothermia can be divided into three stages. At first, body temperature drops two to four degrees below normal, triggering the shiver reflex. Muscle groups around the vital organs begin to twitch and shake in an effort to generate warmth by expending energy. The men will soon be unable to perform complex tasks with their hands, which are becoming numb. The blood vessels in their extremities constrict as their bodies attempt to forestall further heat loss to the air. Breathing becomes quick and shallow. Goosebumps form, raising body hair on end in a futile attempt to create an insulating layer of warm air next to the skin. While this response is helpful for most mammals, it has lost its effect on relatively hairless officers and seamen. Armstrong waves off the other men's concern and presses on.

About six miles from the ship, the party reaches the eastern shore. They climb a hill to plant the English flag—St. George's blood-red cross on a field of white—along with a message in a bottle declaring this "Prince Albert Land" in honour of the queen's German husband, His Royal Highness, the Prince Consort Albert of Saxe-Coburg and Gotha. While the sailors build a cairn to draw attention to this claim, the officers spend two hours climbing a nearby mountain in the hope of gaining the definitive view that will lay to rest any doubt as to whether this field of ice is in fact a strait or a bay. The panorama from fifteen hundred feet above shows the frozen ship, the two small islands, the far shore of Banks Land, and then the hills, gorges, and ravines of Prince Albert Land stretching far behind. Here on the peak, the snow has blown clear, exposing the stones, along with the odd blade of withered grass. All else is white. Never, Armstrong observes, has Nature been more sparing in her gifts.

And yet, to the north, he can almost see the prize. The end of Banks Land is now in view. Like the captain standing beside him, the surgeon takes this as evidence that their ice-choked channel must lead to Barrow Strait, and from Barrow Strait back to England, where they will find themselves swiftly and deservedly acclaimed for having succeeded where so many have failed: they will have discovered and traversed the elusive Northwest Passage. For McClure, however, this hypothesis can only be proved by travelling there himself, by standing on the shore of the strait. He begins to set in motion the plan for an immediate sledge journey to confirm his success.

Still, this view is itself a victory. To celebrate, the men now reach for the slices of bread and butter they have brought along to sustain them. Carried in outside pockets, their food has frozen into solid, inedible lumps. To the cold, hunger is now joined. On their way back down the mountain, they pass fox, hare, caribou, ptarmigan, and lemming tracks in the snow—fresh and abundant proof that, despite appearances, they are not alone.

Back on shore, the seamen have had no better luck. The solitary tin of preserved meat they brought remains uneaten. Upon opening the can, they found the contents so solidly frozen that, despite numerous thrusts of a boarding pike, it could not be dislodged or broken. Instead, they melt snow on their weak cooking lamp and make do with a few mouthfuls of water.

By three o'clock, the party begins retracing its path back to the ship. They cross the floe for about a mile, but soon meet an open channel where the ice has been driven apart by the tide. The black water, 45 degrees warmer than the surrounding air, gives off a disorienting steam. The span across is one hundred feet, and, despite a considerable march along the edge, they find that the farther they travel, the wider the channel grows. Fresh bear polar tracks criss-cross the snow.

Their plan for self-rescue involves breaking off a piece of ice big enough to paddle to the other side. This dangerous plan is quickly scrapped. In the fading light, they instead try to alert the ship. They climb a high hummock and fire off their muskets until their powder is spent. No reply is seen. Having no wood to light a fire, there is now little hope of signalling the crew. They will be forced to pass the night on the ice, an eventuality they had not considered—an emergency for which they neglected to plan.

For the surgeon and other members of the party, the threshold of hypothermia's second stage has been crossed. Their body temperature has dropped between four and eight degrees, shivering becomes more violent, muscle coordination is diminished. Although they appear alert, they wander around in mild confusion. Frequently, they stumble and fall. Their capillaries and blood vessels contract even further as their circulatory systems rally to protect the vital organs. Their faces grow pale. A tingling sensation soon leads to numbness and, for some, intense pain. Feeling is soon lost in their cheeks as the skin begins to freeze. Frostbite occurs not only on their exposed faces, but also on fingers and toes. Soon, the skin will become discoloured and blistered and, if the freezing is not reversed, nerve damage will occur.

The sun has vanished and dark clouds approach. The temperature has fallen to 15 degrees below zero and a breeze pushes in from the south, magnifying the effects of the cold. They share their last ounces of wine, but are unable to melt enough snow for more than a few mouthfuls of water. The wine, while providing the illusion of warmth, only makes matters worse. The alcohol increases the flow of blood to their extremities and skin, thereby accelerating the loss of heat. The momentary flush they feel is their fading thermal energy escaping into the night.

They find it impossible to remain at rest for more than a few minutes at a time. The cold pushes them to stay in constant motion. But in the dark they fall with increasing frequency. Control over their limbs diminishes as their thirst and hunger grow. One of the men revisits that solid block of preserved meat, still frozen fast in its can. When brought in contact with his mouth, his inner lip and tongue instantly freeze and tear.

And then the brother missteps, slips, and plunges into open water. Luckily, he too is grabbed by his comrades and pulled out, soaked and steaming, trousers freezing around his legs and trunk.

While the darkened faces of the men nearby are increasingly difficult to see, the party's predicament is now abundantly clear. They abandon hope that they will be rescued before morning. Each man silently weighs his chances of surviving the night. Some are dizzy and want only to lie down and sleep. McClure recognizes the signs. Everyone must keep moving. The goal now is to find some sheltering hummock where they can all huddle together and await the dawn.

A distant light is seen hoisted from the masthead of the ship. Cannon fire is heard, followed by the sight of rockets streaking across the night. These encouraging signs are soon followed by the flash and report of musket fire between themselves and the ship. The party abandons its search for shelter and cries out in a loud and repeated cheer. At last, their shouts receive a response. Shadows are seen on the far side of the channel, moving in their direction. To the rescue party, the stranded men appear drunk—their speech slow and slurred, their movements unsure.

Some of the rescuers are sent back to fetch a pair of Halkett's boats, a new kind of light, inflatable raft made of rubber cloth. Meanwhile, the stranded men wait on the far side, waving arms, stamping feet, staggering to and fro. When at last the boats arrive, the rescuers lose no time in ferrying their comrades across. But young ice forms on the water and it becomes necessary to break a path with boarding pikes to keep the channel open. Once across, the dehydrated men call out for water. Their rescuers have kept their flasks close to their skin, where the contents remain fluid. They offer up what they have, along with yet more alcohol and a few bites of food.

At half past two in the morning, following eighteen hours' exposure, the party returns to the *Investigator*. They are given a full meal and still more spirits as their core body temperatures steadily recover. All have suffered frostbite—including some of the rescuers—but have been spared the third and final stage of hypothermia. Thankfully, the damage is relatively mild. The returned men toast their saviours, and then dub the high peak from which they viewed their destiny "Mount Adventure."

The following day is set aside for rest, recovery, and the celebration of joining Prince Albert Land to the Empire. Extra food and alcohol are given. Festivities break out on the lower decks. The latest crisis and recovery seem to have bound the crew closer together. Several men have distinguished themselves for their quick and bold action in the rescue, including First Lieutenant William Haswell, who only five months ago was under arrest for insubordination.

McClure retires to his cabin to fill his journal with all that has transpired, and to plan his next move. The brother spends his free time hunched over his desk, adding the new Esquimaux words he has learned to his existing vocabulary. The peace is disrupted only by the news that nearly five hundred pounds of tinned meat have been found spoiled and are declared unfit for consumption.

But there is still land yet unclaimed. A few days on, McClure, Miertsching, Second Lieutenant Cresswell, First Mate Hubert Sainsbury, and Assistant Surgeon Piers take the flag six miles to that pair of small islands that form such a prominent feature in the icescape—those same islands into which they very nearly crashed. These they christen "Princess Royal Islands." The smaller, southern island is only about six hundred yards long and fifty yards wide. Found to contain fossilized seashells and garnets, it has a hill rising a hundred feet above the catastrophe of ice wrecked upon its shore. Inspired by the sheer violence of the scene, Cresswell—the *Investigator*'s twenty-three-year-old artist—stops to make a sketch.

The larger, northern island is about a mile long and six hundred yards wide with a cliff reaching five hundred feet. There, the men discover proof that they are not the first to visit. They locate an ancient cache, along with an ingenious stone trap with an opening wide enough to admit a fox or some other small mammal. The men gather round and consider its design. They reason that once the quarry was lured inside, a large stone—attached to the bait—would fall and crush the unfortunate creature. All are impressed with its engineering. They decide that it must be of great antiquity, judging by the accumulation of moss. Followed by a short survey and a celebratory drink of grog, the Investigators return to the ship, contented with the day's contribution to the Empire's size and glory.

The trap they leave behind was likely built by the Thule, predecessors of the modern Inuit, who dominated the western Arctic from about AD 1000 to 1600 and left evidence of their presence scattered across the land. From kayaks and umiaks, they hunted marine mammals as large as 150-ton bowhead whales—second only to blue whales in mass. They stored food in quantities large enough to sustain permanent winter settlements made of stone, driftwood, whalebone, and turf. On hunting excursions in cold weather, they built domed snow houses; in summer, they used skin tents. Their tools were made of antler, bone, ivory, wood, and metal obtained from natural deposits or through trade from the Norse of Greenland.

After about AD 1000, their territory rapidly expanded eastward across Arctic North America, displacing the pre-existing Dorset culture—the descendants of Paleoeskimos—wherever they went. This eastern migration may have been triggered by a period of warming that reduced Arctic sea ice levels and expanded the available summer feeding area of bowhead whales. Then a period of cooling, between AD 1250 and 1500, brought a mean temperature drop of approximately nine degrees. During this climatic shift, Thule culture went into decline and was eventually replaced with that of the modern Inuit. Bowhead whaling vanished from Inuit culture across Arctic North America and Greenland—with the exception of the north coast of Russian America.

A successful seafaring people, the Thule also hunted terrestrial mammals such as caribou, muskoxen, and Arctic fox and were adept at moving across the land and ice. Perhaps their greatest competitive advantage over the Dorset was the dogsled. Although the Dorset had small sleds that predate the arrival of the Thule, these were likely pulled by people. This technological innovation would have been nothing short of revolutionary, greatly increasing their range and success in hunting and travel. Their techniques were adopted and refined by their successors.

By 1850, the concept of using light dogsleds for snow and ice travel is well known to the British Navy, but they do not adopt it—preferring instead to manhaul heavy sledges, a method abandoned as inferior by indigenous Arctic people centuries before. With few exceptions, expeditionary leaders such as John Franklin and Robert McClure uphold the British tradition. Among them there seems to be a perception that relying on dogs for transportation is somehow uncivilized, undignified, or unmanly. This attitude will persist in Britain's exploration of the Arctic well into the next century—resulting in the occasional catastrophe.

7.

STAR OF PLENTY

All around the *Investigator*, the ice has been levelled and smoothed. The men spend as much time as possible outside, transforming the surrounding hummocks and rubble into a fine promenade and playing fields. But time and light are running short. The gathering darkness will soon restrict and confine. Above the frozen land and sea another paraselene appears—a fearsome trinity of moons.

McClure assembles his team. He enlists Ice Master Court, plus six seamen, to draw the sledge loaded with a tent, blankets, muskets, and provisions to the edge of the strait—nearly seven hundred pounds in all. In a sign of confidence and reconciliation, he leaves First Lieutenant Haswell in charge of the ship. On October 21, 1850, with overcast skies and a temperature of minus-4 degrees, they depart, aided by fifty men carrying their supplies and sledge separately over the rough ice to the end of the rubble field. There, the sledge is loaded and sent off with rousing cheers. McClure's men are further aided by a fatigue party that helps them haul the load fifteen miles before handing over the reins and returning to the ship.

Flat ice and smooth travel are short-lived. McClure and his men get an early lesson in patience and humility from the rugged and broken pack. Pulling the sledge up and down the hard hummocks, they soon crack its frame beyond repair and are forced to send Court back to the ship for a replacement. The remaining men pitch the tent and take their first meal: a pint of tepid water each, fortified with a little oatmeal. Their cooking fuel, spirits of wine, is too weak to bring water to boil. They spread bison skins and blanket bags on the floor of their tent and pass their first full night out on the frozen Polar Sea.

When Court rejoins them the following day, they load the new sledge and proceed north, straining at their harnesses like a team of oxen. After a full day of travel, they are each given another pint of melted snow and a piece of frozen pemmican. They shift and squirm in their blanket bags while the world outside is covered with a fresh accumulation of snow.

At dawn, each man receives his ration of one and a half biscuits and a pint of water, into which a few shavings of chocolate are stirred. After this meagre breakfast, the men break camp, strap into their harnesses, and continue hauling their burden north.

They trudge through dry, knee-deep powder. With few new calories to burn, they grunt and curse as they lay down tracks, perspiration and tobacco juice freezing on faces and beards. One of the men, overcome, falls to his knees and begs for food. While hunger and cold are ever-present burdens, thirst is the growing enemy. But they do not stop and spend precious time and fuel making water. The men are reduced to scooping handfuls of snow into their mouths, further lowering their core body temperatures. Both officers warn against this practice, but even the Ice Master—hidden from view behind the sledge—stuffs his own mouth with snow. By noon, exhausted, they halt for a lunch of frozen pemmican and a second pint of water.

After an advance of some twenty miles, it becomes too dark to continue. They stop and build camp. The cold evening meal is quickly followed by the welcome flavour and aroma of tobacco smoke, each man's weary face momentarily featured in the flash of the match that lights his bowl. Their canvas tent offers a mere eight feet by six of space to house eight men. Inside, they huddle around a candle, repairing worn moccasins and boots. When at last they crawl into their bags and squeeze into position, McClure reaches for his gilt-edged copy of *Chambers's Miscellany of Useful and Entertaining Knowledge*. The heavy tome contains information on biography, history, literature, science, and the arts, stories of wildlife and adventure—in short, everything of interest to the proper Victorian family or party of illiterate seamen. His voice wafts in vaporous clouds. He reads less for their edification than to divert their minds from the bitter cold, their lonely position, and the journey still ahead. At story's end, he pauses to hear their breath fall into shallow and regular patterns. He snuffs out the candle and pulls the blankets around his face and head until only a small passage remains to the cold night air kept dank by the sighs of exhausted men.

They emerge to clear morning skies and a temperature of five degrees above zero. After seven hours straining at their harnesses, and then another equally gruelling day, the goal of reaching the strait seems no closer at hand. The rise and fall of hummocks—interrupted by flat, new ice—continues to the horizon. Although it now appears certain that their channel connects to the strait, the hope of eventually navigating the tumultuous ice next summer seems even more remote. But after sighting and taking a reading on the star Capella, they verify that progress is being made. Having reached 74 degrees, 25 minutes North, they are closing the gap on Parry's furthest reach.

Capella, the brightest star in the constellation Auriga, is in fact a binary pair of yellow giants along with a second, fainter binary. "Little goat" in Latin, Capella is associated with the Greek goddess Amalthea, a foster mother of Zeus. Capella is closer to the North Celestial Pole than any other first-magnitude star, Polaris being only of second magnitude. Capella is thought to define either the left shoulder of the constellation's eponymous charioteer, or the goat he is known to carry. It is a harbinger of plenty; a cornucopia overflowing. With this observation, McClure and his men reckon they are tantalizingly close. They will be the fortunate few to see the old dream come true. Anticipating their near and certain victory, the party crawls into stiff blanket bags, hungry, counting on a few hours' sleep.

The brother climbs down from the ship and onto the ice with plans for an autumn picnic. He sets out for the eastern shore with Clerk-in-Charge Joseph Paine, First Mate Sainsbury, and William Newton, a civilian whaler recruited as a specialist in ice navigation. For this short excursion, they have packed a generous supply of bacon, sardines, and other heavy and high-fat foods their bodies will eagerly metabolize into both heat and energy. The brother also carries the rifle given him by McClure. After a two-hour hike over rough ice, the men arrive at a beach strewn with driftwood. Soon, this fuels a roaring bonfire and boiling kettle of coffee. After a warm and filling lunch, they set out in search of ptarmigan. The only animal to cross their path, however, is a startled Arctic hare—which they dispatch with ease.

The cold chases the hunters back to the site of their fire. Soon, they see distant figures approach through the heat shimmer and smoke. Observing the shapes for a time, they conclude it to be a party of Esquimaux. Amazed with his luck, the brother begins making plans to spend the long, dark winter with these folk, informing them of the ways and means of the Almighty while expanding his Esquimaux vocabulary. It seems God's plans for him in this frozen place are at last being revealed. He awaits the appropriate moment to signal greetings of friendship and peace. All four Investigators gaze in expectation as the visitors continue to draw near. But it soon becomes clear that this is no family of Esquimaux, or men of any kind. These are five large, dark beasts. Black and brown bears are not found at this high latitude. The men quickly set about trading birdshot for balls. Sainsbury, the frailest member of the group, is unable to load his gun—his frostbitten fingers have gone numb.

The men lay out flat in the snow twenty feet apart, sighting their targets as the beasts close in. Despite the fire, their presence appears to have gone unnoticed. And then, at sixty paces, the beasts suddenly halt. They snort, stomp their forelimbs, and huddle together with lowered heads facing the men and their guns. The largest member of the group slowly approaches, while the rest remain motionless behind. These, Miertsching realizes, must be those creatures described by the Arctic explorer Sir John Ross in his *Narrative of a Second Voyage in Search of a Northwest Passage.*

At first glance, they resemble shaggy bison. Upon closer inspection, they appear to be made of constituent parts of bison, caribou, and Scottish highland cattle—with a loose yak's pelt thrown over their backs for added bulk and warmth. Muskoxen, in fact, are more closely related to goats. This small group consists of only five individuals. There is, of course, safety in numbers, and muskoxen ordinarily travel in herds of ten to twenty, occasionally in groups of seventy or more. On a diet of grasses, sedges, and willows, they can grow eight feet long and shoulder-high to an Englishman. The largest weigh well over eight hundred pounds. Their rich, voluminous fur drapes down to the snow. Almost hidden in the wool near their eyes is a small scent gland that produces a strong, musky odour. Both cows and bulls possess impressive, low-slung horns, which curl broadly over their foreheads and then down into sharp points like handlebar mustaches. These are used in displays of male dominance throughout the year, particularly during the summer rut. Their dramatic contests begin with threat displays, in which bulls rub their scent glands against their forelegs or the ground. Next, contestants slowly back away, swinging their heads, then charge forward, colliding head-on with a thunderous crack—the force of impact absorbed by their thick horn ridge and skull.

Early European accounts of muskoxen note their distinctive defensive posture. When attacked by wolves or human beings, the herd runs to an area of shallow snow or high ground to line up in a wall. Once surrounded, they press their rumps together, forming a tight ring, enclosing vulnerable calves in the centre. From this position, both bulls and cows take turns lunging at their tormentors. For the Inuit, they are a critical source of meat, warm sleeping robes, and horn—used in the manufacture of weapons and other implements. Their word for the beast is *omingmak*, "the animal with skin like a beard." Its ability to survive gale-force winds in extreme sub-zero temperatures is due largely to the insulating properties of their remarkable, shaggy coats. Their wool, or *kiviut*, is stronger than sheep's wool, eight times warmer, finer even than cashmere.

A contemporary of the mammoth, muskoxen (or their ancestors) likely migrated to North America between 200,000 and 90,000 years ago. During the last ice age, they seem to have survived on isolated areas free of glacier ice. As the glaciers retreated, they gradually migrated across North America to Greenland. The last Old World population of muskoxen died out in northern Siberia about 2,000 years ago. In Arctic North America, their last refuge, the arrival of Europeans and their guns has tipped the balance of power. Countless muskoxen have already been shot for their hides by commercial hunters, or killed to feed whalers and explorers.

The advancing bull reaches to within thirty paces of the hunters and stops, lowers his head, and stamps the ground with his hooves. He brazenly tosses snow into the air with his horns and snout. The first ball passes through his head, about four inches above his nostrils. As he turns in panic, the second ball enters his ribs. He rushes back to the herd and then turns his bleeding face toward his attackers once again.

The men crawl on elbows and knees and then open fire with a volley from three double-barrelled guns. The beasts, now panicked and enraged, charge as the men rush to reload. Soon, three muskoxen lie dead while another rushes directly at Miertsching. The brother again takes aim, holds his breath, and squeezes the trigger. But the priming falls and the gun misfires. To avoid the flailing horns, he jumps aside, trips, and sprawls out flat. The frightened muskox rushes past and gallops on, leaving a bright trail of blood in the snow. The brother scrambles to his feet, replaces the priming, and follows the tracks.

His quarry is soon found standing in deep snow, panting, bleeding from his wounds. The brother shoulders his gun, fires, and at last the animal falls. He then turns and runs back to help the others, but he is too late. The final beast is laid out on its side, muscles atwitch, eyes dulling over as stiffened legs go limp. Five muskoxen now lie dead and the hunters congratulate one another on their startling success. Their joy overwhelms them; only last week another seven hundred pounds of preserved meat were pulled from the ship's stores and declared unfit for human consumption. They stand over their kill smiling, admiring their size and quality, thanking God for their good fortune, imagining the fresh steaks and stews they will enjoy over the long polar night.

But First Mate Sainsbury and the whaler Newton are now in danger. Their frostbitten fingers require an immediate return to the ship. The brother and Clerk-in-Charge Paine remain to guard their prize from wolves and bears, whose tracks and signs abound. They kindle their fire, sip cocoa, and await the speedy arrival of a sledge bearing a tent, blankets, food, and six fresh companions. They make camp and help collect the carcasses, hauling them together where they can be guarded throughout the night.

The men stretch out inside the small tent, alternating heads and feet, reminding the brother of herring in a tin. A lantern is hung from the post, illuminating their corporal steam. Fog accumulates to such an extent that it is difficult to see from one end of the tent to the other. And then the light's snuffed out. Sharing a series of blanket bags, the men are packed so tightly it is almost impossible to move. Paine is in the middle. He has developed a cramp in his leg and repeatedly digs his heel into his neighbour's back. Miertsching endures this irritation until he can stand no more. The mild, pious brother suddenly erupts in loud and surprising complaint, drawing peals of laughter from his companions. Following a series of rude jokes and innuendo, the stillness of the night prevails.

By morning, the tent's interior is covered with frost. It is as if their breath has turned to snow. The foul air is soon relieved with the scent of a driftwood fire. The brother crawls over late risers, steps outside, and is handed a mug of cocoa. Someone slices into the abdomen of one of the bulls, and soon a rich breakfast of liver, kidney, and heart darkens and drips over the flames.

That afternoon, they return to the ship to find the entire crew out cutting a path through the ice. The Investigators receive their hunters with enthusiastic cheers. Everyone wants to examine the beasts now laid out near the hull. At the centre of the excitement stand the brother and his fellow hunters—two of whom will be recovering from frostbite for weeks to come. The flesh of four bulls and one cow is weighed: 1,296 pounds of muskox meat (excluding heads) are added to the ship's larder. The brother enjoys telling and retelling his story, smiling at the thought of McClure's reaction to this bounty when—God willing—he and his party return.

The early light is lavender, with an eastern blush of pink. The air, cloudless and clear. The morning of October 26, Robert McClure and his men quit camp before dawn and trudge to a nearby hill to gain a better view of the frozen sea. There, six hundred feet above, they await the rising sun. Along with illuminating the chaos of ice, its light soon strikes the visible end of Banks Land—now clearly Banks Island. From the perspective McClure and his men command, it is apparent that there is no land between them and Melville Island to the northeast. Without a doubt, this is the land sighted by Parry thirty years before. After discussing the various features of the icescape before them, the men gaze on in silence.

"Thank God," McClure says for all.

In the aching cold, they turn and congratulate one another on their discovery of the Northwest Passage. Centuries of dreams and schemes have once and for all been proved. It is one thing to reason, to calculate on a chart or globe—quite another to see it through frosted lashes. But when the frost is wiped away, the unfinished business remains.

Has Franklin already seen this passage? Is he now making his way back to London's comforts and praise? Here, on such a promontory, there is no marker of any kind, not the slightest evidence of his passing. The best they can do is savour the moment for themselves and all those who have suffered in the attempt to reach it. But the remembered risks of having achieved this view are quickly forgotten. Now it must be crossed. If McClure can somehow manage to sail his ship past this point next summer, he will complete the first circumnavigation of the Americas and traverse of the Northwest Passage. His name will take its place among the greatest navigators the world has ever known.

They make camp on shore. With pieces of broken sledge and stems of Arctic willow, they build a humble fire to mark the occasion. To toast their victory, McClure supplies each man with a cup of grog. But the temperature continues to fall. Their beards have frozen into masks, their bison skins form a solid mass, and their rigid blanket bags can be stood on end. They must lie upon their bedding for hours before it is pliable enough to use. But sleep will not refresh these men; an enduring fatigue sets in.

The following day, they erect a large cairn containing a proclamation of their achievement and the position of their ship. Then, in the face of a sharp and biting wind, they lean into their harnesses and drag their sledge toward home.

On the morning of October 30, the Princess Royal Islands come into view. McClure leaves the party in the hope of reaching the ship early so relief can be sent for his weary men. Calculating that he will reach the *Investigator* in four hours, he sets out alone across ice fields, marching up and down the hummocks and rubble. But the day runs short and the darkness rushes in. Mist rolls down from the strait, and then snow starts to fall. The ship is still six miles away. Unable to see his path, McClure presses on, choosing his course by wind direction alone. But he trips and falls amid the upended floes, tumbling through sharp-edged blocks and new-fallen snow. Now the possibility of a broken neck or leg is added to the fear of freezing.

McClure climbs a high hummock in the hope of gaining a glimpse of his ship or sledge party. He fires one of his two loaded barrels in an attempt to draw attention. For an hour he scans the obscured horizon, searching for some sign, until at last he is rewarded with a dim blue glow cutting through the falling snow. The sledge team, he concludes. He immediately fires his second round to signal his position. But this receives no reply. Both chambers are now spent, and he has packed no spare ammunition. And then the blue light fades.

Sometime later, the light glows again, but now seems farther than before. When it fades this time around, McClure is left with the realization that he must endure this night alone. He recalls the two distant polar bears he saw earlier this day and grips his empty gun.

Two hours pass; twelve hours of darkness remain. The temperature falls to fifteen below zero. He has brought nothing to eat or drink, but remembers the single match in his pocket. With stiff fingers, he pulls it out in the hope of reading his compass. But the match flashes out in an instant, returning him to darkness.

Perhaps the sledge party will reach the ship and send relief. This thought encourages him to pace back and forth on his high hummock for another two hours until this hope slips away.

The cold envelops and pervades. While hunger continues to gnaw from within, it is thirst that will drive him mad. Now, despite his earlier warnings to his men, he fills his mouth with snow. Recognizing the signs and stages of exposure, McClure is aware that his judgment is failing. He is neither a strong nor robust man. Until now it has been his judgment—and force of will—that has always seen him through. He has come too far to let this single night defeat him. But weariness overtakes conviction. Under a thick ledge of ice, he finds a bed of snow and lies down to sleep.

A few hours later, McClure opens his eyes to bright, orderly constellations and the flash of a passing rocket flare. Is this a dream? He struggles to his feet to see that the sky has cleared and the aurora is lighting up the night. But the silhouettes of the Princess Royal Islands and the *Investigator* remain unseen. His body and mind begin to disregard his commands. Core temperature drops dangerously low and the shiver reflex begins to fail. Thoughts are sluggish, muscle coordination diminished, limbs clumsy and difficult to control. His pulse and respiration decrease, but his heart beats faster.

McClure stumbles upon his own track in morning's first light. He finds the presence of mind to retrace his steps until he sights a far black mass—a view that brings both mortification and relief.

Word comes down that the men on deck have spotted a lone figure approaching the vessel. Miertsching joins his fellow officers in the rush up top. By the time he leans over the bulwark, the thin shadow has nearly reached the hull. It is the captain, minus his sledge party, moving like a man in shock, less than half alive. Seamen rush down to offer aid, then hoist him up to waiting hands. They inundate McClure with questions but receive no reply; only a slow, vacant expression and unsteady, jerking movements. His limbs are stiff and weak, he is unable to shape a word. The surgeon and his assistant take command of the situation and order the patient carried to his cabin below.

The wait for word of the whereabouts of the missing party is relieved three hours later, when seven men are seen approaching the vessel. All seemingly alive and ambulatory, they call out for news of their missing captain. When assured that he is safely aboard, they shrug as if a great burden has been lifted. All available hands turn out on the ice to welcome them back and congratulate them on their success. They rush forward to grab the reins and relieve their weary comrades. Once comfortably aboard, Ice Master Court and his men relive the sweep of their ordeal and discovery.

Assistant Surgeon Piers examines the returned men and finds that, while resilient, they have suffered greatly. All are thin and wasted, frostbitten, and will require a considerable term of rest and recovery. Over the course of their journey, they crossed 152 miles of ice, heavily burdened, at the absolute limit of the season. This feat was accomplished on the following average daily ration:

1/2 pint cocoa, tepid
7 ounces biscuit
4 ounces pemmican
1 ounce oatmeal
1 gill (5 ounces) rum
3/4 pint water

Their sledge is brought aboard and studied like a precious artifact. Despite the consumption of provisions and fuel, it is found to weigh 793 pounds—a full 100 pounds *more* than when it left nine days ago. The entire increase, the officers deduce, is a result of the accumulation of ice and frost condensed from the men's breath and perspiration upon their clothes, tent, bison skins, and bedding.

Two days later, the ship's company is called to muster on the outer deck. Captain McClure—weak, chastened, but now again able to speak—is determined to make a good impression. He praises the men for their conduct and implores them to maintain their cheerfulness and camaraderie as they embark on the long polar night. This, and a rigorous exercise regime, will be critical in maintaining both physical and mental health. But today, he says, everyone should take a moment to recall the dangers faced, the accomplishments achieved—and savour the success together.

"It is my genuine hope," McClure says, "that by this same time next year, we will be approaching the shores of Old England to spend that next winter among family and friends. And I promise that when we return, you will not be forgotten. I will recommend the crew of the *Investigator* to the Admiralty as favourably as I know how . . . Please enjoy tonight's feast in celebration, and with my sincerest thanks."

The men respond with three cheers each for the Queen, the discovery of the Northwest Passage, Captain McClure, the rest of the officers, and absent wives and sweethearts.

Over the following week, the temperature continues to slide. The new ice is now twenty inches thick. The captain, slow to recover, is forced back to bed with a body covered in sores. Among the crew, incidences of frostbite increase—the surgeon himself falls victim to a severe case in which his right hand is frozen into an inflexible hook. He will lose the use of it for two months to come.

The *Investigator* is made ready to face the Arctic winter. Hatches are closed and a housing placed over the ship. A sixteen-inch layer of "polar cement"—snow, sand, and gravel—is laid on the upper deck. With saws and cutlasses, blocks of snow are cut and stacked around the hull until they form an insulating wall, eight feet thick. On either side of the vessel, a stairway is cut from the deck to the surface ten feet below. The icescape around the *Investigator* is carefully smoothed and levelled so that no one will stumble in the dark.

Copper pipes are passed through the decks and out into the open. These allow the escape of the foul air that pools inside. The ventilating pipes, along with the daily evacuation of the lower decks during exercise, will allow for circulation. Throughout the vessel, fires are lit: the Sylvester stove to provide general heat, the galley stove for cooking, and a small stove each in sickbay, the gunroom, and the captain's cabin. Coal rations are calculated and set.

On November 11, the Investigators clamber up on deck to say goodbye. It is a still, clear, day of minus-26 degrees. When the sun grants its brief midday appearance—the last of the year—the men brighten as in the presence of a lover. And when it dips back below the horizon a moment later, they linger, lavishing praise on the colour and beauty that yet remain.

8.

ENDURING NIGHT

Time, a human construct, has its way with us in the end. Being able to mark its progress gives us at least the illusion of control. Remove the pre-eminent indicator by which it is measured—the passage of the sun through the sky—and men are easily cast adrift. Undeterred, time moves through the gloom. Routine is offered as a tool to tame and bend it to our will or, at the very least, keep it from driving us mad. This is, perhaps, the best hope for officers and crew idled together on a dark and frozen sea.

At 5 a.m., the night watch gives his signal. Candles and lamps are lit—bright blooms in the otherwise impenetrable black. On the lower deck, sailors tumble out of hammocks, rubbing sleep from their eyes. Their still-warm beds are unhung, rolled, lashed together, and piled against the bulkhead like abandoned cocoons. The ship is then inspected, swept, and put in working order for the day. The officers and captain hear this reveille, but turn over in their berths to luxuriate in two more hours of sleep.

At eight o'clock, the line forms in the galley. Steam from the cauldron and kettle collides with the cold air following the men up from their quarters. Assigned to mess in groups of eight, sailors trade the day's first insults, witticisms, plans, and dreams over the breakfast ration: one cup of cocoa, half a pound of salt pork. Soon the order is given to don outer garments. A few men remain behind to clear tables, wash dishes, and scrub the galley; the rest proceed up the ladder to the upper deck, emerging from the open hatch in a dramatic cloud of steam. In the space of just a few short steps, they pass from a bright, humid space of 50 degrees above zero to darkness that plunges 30 degrees below.

At nine o'clock, they receive the signal to muster, falling into rank and file for inspection. All accounted for, they are ordered off the ship and onto the ice. Six of the men set out to gather fresh snow to be melted down by the cook; the rest are left to their own designs.

At twilight's first blush, they divide into teams for rounders, a game played with a hard, leather-covered ball, a bat, four bases, and teams that alternate between batting and fielding. The batting team scores points by completing a circuit around the bases. Keen competition ensues. A version of the British game, known as baseball, is popular in New England. Rounders is often followed by football, a game in which most everyone is kept in motion. Walking, wrestling, and playing catch are also popular pastimes. At half past eleven, the men are called back onboard and given a dose of lime juice to forestall the appearance of scurvy.

Dinner, the midday meal, consists of either pork or salt beef, accompanied by dumplings, peas, and dried cabbage. At half past one, the crew is sent back to the ice to march around the ship, where they discuss life histories, gossip, future prospects, and schemes. Officers are served dinner at two o'clock and take their exercise as they see fit. At 4 p.m., the crew is signalled. They stumble aboard, shivering, to a supper of more salt pork and black tea. The master-at-arms issues the daily ration of grog, which is customarily enjoyed with a pipe of tobacco.

At long last, the promise of establishing an academy for sailors is fulfilled. School is held throughout the winter to teach the men to read, write, and make simple calculations. Clerk-in-Charge Joseph Paine is drafted to the role of chief instructor for the entire academy. The other officers, who are by and large at leisure, do not offer their assistance. The exception is Ice Master Stephen Court, who holds a course in navigation. Miertsching pokes his head in to observe the big, bearded seamen with pens in their fists, wrinkling their brows over leaves of paper as they form letters and numerals—skills he himself acquired as a child. Their concentration is inspiring. The brother notes that the foul language so prevalent earlier in the voyage is now rare among these students. Classes run between six o'clock and half past seven, frequently followed by an hour of poetry recitations, music, song, and dance.

A popular dance is the "railway," which features the fiddle and stomping dancers mimicking the sound of a locomotive building a head of steam, chugging up and speeding down hills, and then—after a considerable trek—finally pulling into station. Assistant Surgeon Piers is in the habit of pacing the outer deck after his evening meal. He is especially fond of the rhythmic thumping and whistling below. It transports him back to England, where he imagines himself gazing out the window of a train at the lush, varied, and brightly lit countryside passing by. Piers is pleased to find the men cheerful and content as the full weight of winter descends. Their mood is especially light at times like these, when reminded of home, where all expect to be this time next year, captivating loved ones with tales of their successful discovery of the Northwest Passage and survival of a winter at the top of the world.

While on the whole the men are healthy, Ship's Surgeon Armstrong is kept busy treating blisters formed on frost-bitten faces, fingers, and toes. Overly calculating and quick to share his opinions, Armstrong finds his company politely tolerated by fellow officers at leisure time. For relief, he occasionally wanders down to sit with the crew in their low, cramped space, where the air is close and thick with smoke. Following school and their more formal recreations, the men retire to their quarters to spend the last hours of candlelight playing checkers and cards, or employed in their various sub-trades as tailors, knitters, or cobblers. Those who have no particular craft or skill busy themselves with needlework. The surgeon is impressed with their diligence. The men quietly mend and repair as their "scholar"—a literate seaman—reads aloud some gripping tale until the hammocks are re-hung, the lights snuffed out, and the endless cycle resets and repeats.

In a world seemingly sterile and bereft, the exotic sounds and scents of the *Investigator* begin to draw attention. From low on the horizon, a pair of common ravens appear. They fly over the snow-covered landscape and frozen sea toward the ship, wheeling several hundred feet above as the men march around their grey-brown circle below. The ravens' mournful croaks are often the only native sound heard throughout the day. At first viewed as dark omens, the sight and sound of these familiar birds is soon eagerly anticipated, relished, and then discussed at length. These visits make the men feel less alone.

Explorers and naturalists have long been aware that common ravens are accomplished adaptors, surviving Arctic, temperate, and desert environments. These intelligent opportunists are encountered across North America, Eurasia, Central America, and northern Africa, making them one of the most widespread species on Earth. The largest of the songbirds, ravens have coexisted with people for many thousands of years, leaving deep and lasting cultural impressions wherever they are found, including England, where King Arthur himself is said to have escaped the death of mortal men by transforming into one. At the Tower of London, they are kept as mascots. According to tradition, they have resided there since at least the reign of Charles II (1660–85), who was warned that if the birds were dispatched to make way for a new royal observatory, ruin would befall both Crown and country. It is said the king decreed that at least six ravens be kept at the tower in perpetuity.

Ravens are known to mate for life, and the Investigators come to look upon this couple as both friends and neighbours. After a period of weeks, however, only one of the pair returns. This ignites a round of speculation as to the whereabouts or fate of the absent bird. The consensus is that she has fallen victim to a fox. When the widower alights on deck, he is given food before flapping off again.

It isn't long before the Investigators awaken one morning to find a vixen caught in a trap near the ship. The men crowd in for a look at this new distraction. A year-old Arctic fox waits silently inside her pen, staring back at the gawkers with black, liquid eyes, making no move to cower or escape.

Arctic foxes are found throughout circumpolar lands. Unlike the wide-ranging raven, this is their only home. Compact and wily, they weigh between six and twenty pounds and are generally less than a yard long from snout to the tip of their tail. Related to the familiar and somewhat larger red fox hunted for sport by English noblemen, Arctic foxes have a short muzzle and ears, and fur on the soles of their paws. At this time of year, they trot across the ice and snow with hurried determination.

Although this specimen appears delicate and slight, she is extremely hardy, able to survive and thrive exposed to frigid extremes. Her luxuriant coat is the absolute white of her surroundings, rendering her practically invisible until she moves. With the arrival spring, her fur will take on the brown-grey tones of rock and tundra. This highly effective seasonal camouflage helps her stalk and hunt lemmings, voles, birds (and eggs), ringed seal pups, and even fish. In winter, much of her time is spent faithfully shadowing polar bears, taking care to keep a respectful distance, ready at a moment's notice to scavenge or snatch leftover morsels of seal.

Given the chance, she will likely mate next season. In times of plenty, Arctic foxes are incredibly fertile. With an average litter size of eleven whelps, they can produce up to twenty-two offspring in a single litter—the largest known for any wild mammal in the world. They raise young in sprawling dens up to three centuries old and with as many as a hundred holes for speedy entrance and escape. An average family of Arctic foxes can consume up to a hundred lemmings each day, and fox populations fluctuate dramatically with the rodent's cyclical boom and bust. This highly nutritious prey is captured with a quick dash and cat-like pounce. Like other Arctic foxes, this vixen will be an attentive mother for three or four months. Then, as fall and winter approach, her whelps will scatter and her coat will turn white once again.

The men decide to adopt and tame her. However, Mongo will have none of it. More than twice her size, the ship's dog is afraid to face the vixen. The men are ashamed for him and the rest of his breed. A week later, the fox escapes with her course in domestication only just begun.

Additional traps are set. The next three foxes caught are not indulged; they are slaughtered for their luxurious white pelts, each of which is claimed as a stole for an absent wife or sweetheart. Mongo, encouraged and emboldened by his masters, evolves into something of an English foxhound. When yet another victim is trapped, the men spread out in an enormous circle to prevent its escape. The prisoner is let loose in the centre and Mongo pursues it for a time, until the dog tires and the cold forces the men back inside, leaving the exhausted fox to survive another day.

On November 30, four weeks past his return and triumphant address, McClure makes an appearance on deck looking wan and wasted—although reportedly on the mend. He takes a tentative stroll around the ship, observed by the officers and crew. The brother, suffering rheumatic pain brought on by the damp and cold, also takes to circling the ship for exercise, then marching further across the unending ice. In the dim twilight, there is no feature to catch his eye, no rock, bush, tree, or building to mark the canvas of snow. The still air is so devoid of sound he can hear the tick of his pocket watch. Those without watches claim they can hear the beat of their heart.

On the surface, it appears as if nature has perished. And yet the sky offers numerous diversions. The flashing constellations, visible night and day, are often crossed by shooting stars. A halo of coloured light can be seen encircling the moon. On December 21, the shortest day of the year, the stars are so bright in the moonless sky it seems possible to read by their light. The brother tests the cliché by bringing several books out to the ice. He manages to read the one with the largest type.

Christmas Eve is a relatively mild minus-15 degrees. The galley is decorated and the larder raided for a banquet befitting the occasion. The freshest frozen beef is served, along with sirloin of muskox, mincemeat, and Scottish sweets—all washed down with extra rations of grog. The officers and captain enjoy a quiet dinner including these delicacies and more. And then pandemonium ensues. To Miertsching, the debauchery overtaking the ship is both a shock and offence. There is no hallelujah, no hymn or hosanna. The men seemingly lose their senses, shouting and singing rude and inappropriate words as loudly as they can manage. The brother cannot imagine a more unholy Christmas Eve. In an act of defiance and absolution, he bundles up and leaves the ship to offer prayers, meditation, and thanksgiving appropriate to the occasion.

Praise to the Lord Jesus for His grace, mercy, and truth. In all these trying and disagreeable circumstances, surrounded by enemies of the gospel and, above all, missionaries, I will yet pass comforting and holy hours . . .

The brother remains outside on the ice, alone, until he can stand the cold no more. Exhausted, chilled to the core, he stumbles back aboard and down to his cabin, where he shuts the door and burrows into his berth. As the party rages on, the men become so thoroughly soused that two officers are forced to keep watch to prevent anyone from starting a fire or wandering away from the ship to certain death out on the ice. The wanton consumption of alcohol recalls that night the crew broke into the rum to drown their fear in the face of catastrophe. This night, this Holy Night, there will be little chance for sleep.

At 9 a.m. on Christmas morning, the men are mustered on deck for inspection. All are met with the grave expression of McClure, who was forced to endure a sleepless night. Ordinarily indulgent, even fond, of "spirited" behaviour, this day—this Christmas Day—the captain is not amused. He announces that there will be no divine service today, no prayers to mark the birth of the Saviour, because too many men are still drunk. Being forced to issue such an order is both a disgrace and shame.

Near the ship, a modest hole is kept open in the ice—now thirty inches thick. This ensures a ready supply of water should fire break out again. Hourly attention is required to keep it from freezing over. The sailor sent down to perform this task is startled to find a blunt head and two round eyes staring back at him. A ringed seal has surfaced to take a breath and a quick look around. Shaped like a fat, poorly rolled cigar, the intruder closes his nostrils and drops away silently, vertically, keeping watch all the way down in case the man decides to attack or follow.

Reaching only about four and a half feet in length, and weighing between ninety and two hundred pounds, ringed seals are the smallest and most common seals encountered in the Arctic. Their distinctive, mottled fur appears to have been splattered with bleach. Preferring pressure ridges, polynyas, and leads, they are the only species of seal capable of maintaining a breathing hole in ice, which they excavate using the claws of their flippers. They are completely dependent on polar ice, which provides platforms for hauling out to rest, giving birth, and rearing pups.

Along the edge of the ice, they dive to depths of three hundred feet in search of a variety of fish and crustaceans, remaining submerged for up to forty-five minutes at a time. Before surfacing, they often blow bubbles at their breathing holes to draw out any predators lurking above. In early spring, females give birth to a single pup in a natural snow cave or excavated lair. Ringed seal pups are nursed for about two months, after which they are abandoned. Roughly one in four are captured and consumed in their dens by Arctic foxes. Both pups and adults are the primary food of polar bears, which eat about one a week. They are also preyed upon by walruses, orcas, wolves, ravens, and men.

The Inuit have several names for ringed seals, depending on a creature's sex and age. They hunt them at breathing holes with extraordinary patience and ivory-tipped harpoons. When pursued, ringed seals dive and swim under the ice between a network of breathing holes. The lucky, careful, and wise can survive for over forty years. From a seal's perspective, the hole next to the large, dark belly of the *Investigator* should be clearly suspect. However, curiosity often gets the better of them.

On the final day of the year, the mercury dips to minus-40 degrees—the coldest thus far. The men sing and dance after supper, but without any extra measure of grog their celebration is subdued.

New Year's Day, 1851, arrives with a spirit of reconciliation. Festivities begin with extra rations and grog at breakfast. At morning muster, McClure encourages the men to enjoy themselves in a more orderly and dignified manner so as to enter this new year truly refreshed and redeemed. Following this speech, the captain leads the entire company down to the seamen's quarters, which Miertsching is surprised to find festooned with flags, coloured paper painted with anchors, and romantic pictures of the sea. These decorations surround bright tables arranged in elegant and civilized settings. The officers are greeted by the clean and smiling faces of smartly dressed sailors, who give their guests three welcoming cheers. The brother cannot mask his surprise; he is overcome. It is as if the men acknowledge their past transgressions and now wish to make amends. This, he believes, is the celebration that should have greeted the Lord's nativity.

BILL OF FARE

1st course

turtle soup, preserved salmon & eels

2nd course

forequarter of fresh mutton, stewed muskox,
beefsteak pie, York ham, curried salmon,
lobster patties, green peas, carrots, parsnips,
potatoes, Indian & English pickles

3rd course

plum pudding, mince pies, gooseberry tart,
gooseberry jam tart, damson tart, raspberry
creams, jellies, porter, ale & sherry

dessert

frosted plum cake, preserved China ginger,
almonds, raisins, citron, preserved gingerbread,
nuts, sweet biscuits, port & sherry

Following the banquet and coffee, the men invite their guests to stay for a theatrical presentation. One sailor, dressed as Lord Nelson, duels with another dressed as an admiral of the French navy. They battle over Alexandria and Gibraltar. Everyone cheers the dramatic and foregone conclusion. This is followed by a selection of sea songs, and then everyone stands for "God Save the Queen." There is no loud, drunken excess—only the call to hammocks. The brother thanks his cordial hosts and retires to his cabin smiling, content with the display of goodwill, fraternity, and the seeming restoration of order.

9.

THE SEARCH

Measured in five locations, the young ice now averages three feet, eight inches thick. As it continues to grow, the frequency of pops and squeaks builds until it culminates in the sound of musket or cannon fire. Occasionally, a far-off crack mimics thunder. Inside the ship, corporal and cooking steam rise from the lower decks, producing great quantities of frost on the beams and floor in the officers' cabins. Between three and four buckets are collected and disposed of daily. This task is woven into the established routine of sports, meals, school, music, the reading and rereading of books.

Alexander Armstrong busies himself by examining the ship's company at the beginning of each month. This provides an opportunity to stave off the first signs of scurvy. The men are pallid, but in otherwise robust health and spirits. There are only two patients in sickbay: one with a bowel affliction, the other suffering from a serious heart condition. This second patient is made as comfortable as possible, but he drifts in and out of a critical state. The prognosis is one of guarded optimism.

With seemingly limitless hours to fill, Miertsching relines his sealskin coat and makes felt boots for both himself and Captain McClure. The two men often walk together, circling the ship, the brother offering unsought spiritual council. McClure finds his mind wandering across the Arctic, wondering just how close their nearest comrades might be.

The Investigators are aware of various parties involved in the search for the Franklin expedition, amounting to perhaps one of the greatest rescue efforts ever known. Sir John Richardson and Dr. John Rae may still be somewhere south and east on the mainland coast, as well as W.J.S. Pullen leading both the *Plover* and the *Herald*. Farther east, William Penny leads a search with the *Lady Franklin* and *Sophia*; Horatio T. Austin commands a four-ship Admiralty expedition; Charles Forsyth and Sir John Ross each head a privately financed expedition; while Edwin J. De Haven commands the first U.S. expedition in support of the rescue effort. Here, at the far western edge, the whereabouts of the *Investigator*'s consort, the *Enterprise*, remains a mystery. Perhaps some of these expeditions are as lost as the object of their search. It is a safe bet that none of the other rescuers have solved the mystery of the Passage. Undoubtedly, whoever remains on the edge of the Polar Sea is similarly seized, counting hours and rations in wait of the pack's dissolution.

One morning, seaman and sickbay assistant John Ames is out walking the darkened ice. There, he spots something no one expects to see: a small herd of caribou marching through the snow toward the Princess Royal Islands. He is able to creep up on the animals, approaching to within twenty yards. Like everyone else, Ames was convinced that most land animals had migrated south for the winter. This was a belief passed down from both the eminent naturalist and explorer Sir John Richardson and the late admiral Sir Edward Parry. Ames quickly returns to the *Investigator* to inform the captain, who triumphantly records the event in his journal. With this new discovery, his expedition now seems to have rewritten natural as well as geographic history.

By January 6, 1851, the darkest stretch of winter is behind, but the coldest season remains. Each day, the twilight gains confidence. On January 12, the mercury freezes in the thermometer at minus-45. For the next two days, the temperature falls farther still, to an estimated minus-51. Regardless, the men are unwilling to give up rounders. Among the players, minor frostbite and a dislocated shoulder are the only injuries. Mongo, the ship's dog, never shows the slightest hint of discomfort from the cold. Despite numerous invitations, he does not descend with the men to the warmed decks below. At night, he shelters in his kennel on the upper deck. When the men leave the ship, he accompanies them on walks and to the rounders pitch, where he curls up and falls asleep directly on the frozen sea.

Another seventy-four pounds of provisions are declared unfit for human consumption. The men's loss is Mongo's gain. The dog is presented with an eight-pound tin of spoiled meat. He gorges himself, then carries the leftovers away from the ship. He keeps careful watch on these treasures, often patrolling his circuit of caches. Whoever dares walk in their direction is followed close behind. Venture too near and the dog gently takes hold of a man's boot or trouser leg and gives it a firm yank.

An indefinite black speck is seen low on the horizon. This sight is greeted with joy. The raven—christened "Ralpho" in absentia—is on his old flight path to the ship, and he is not alone. The men gather on deck to greet him and his new mate. After such a long absence, his return has a profound and positive effect on both the officers and crew. They have come to see their fate bound up with this bird. Clearly, this sign foretells their own survival of winter, return to England, and resumption of full and productive lives.

On February 3, the men gather in anticipation of another reunion. After having abandoned them to darkness and twilight for eighty-four days, the sun at last returns. At 11 a.m., it reaches over the eastern hills through a rent in the clouds to meet their frozen beards and lashes. The Investigators let out a spontaneous cheer. They turn and congratulate one another on their perseverance, then chase their long-lost shadows around the ice and snow. They vow to never again pass such a bitter, interminable night.

Two men return from the Princess Royal Islands and report having been followed by a wolf. This news is not taken lightly. The crew arm themselves and set out like vigilantes—but only tracks are found. After they return, the beast is spotted near the hull. Marksmen take up defensive positions.

It appears only in flashes, ghost-white from a distance, greyer tones observed on the chest and abdomen. Like most other mammals found on the shores of the Polar Sea, the wolf has evolved a winter disguise. Cloaked in extremely dense under-fur to insulate against the cold, they have narrow frames, powerful backs, and long legs for loping across vast tracks of ice, tundra, and snow. Their bodies are covered in scent glands, which are used for marking and navigation. They can stretch to over six feet in length, snout to tail, and weigh up to 175 pounds. Although this animal appears alone, Arctic wolves ordinarily hunt in packs, which enable them to pull down caribou and muskoxen—animals up to eight times their individual weight. Together, they will roam over as much as a thousand square miles in search of their principal prey, which they subdue by tearing at their haunches, causing a massive loss of blood. Occasionally, they go for the throat, biting through the windpipe or jugular vein. They supplement their diet with hares, seals, ptarmigan, lemmings, and birds. Highly social and aggressively territorial, they will attack and kill competitors.

Arctic wolves are a subspecies of the grey wolf, the world's most widely distributed carnivore, with a range that once included Europe, Asia, North America, and northern Africa. Wolves have proven themselves capable of adapting to most types of landscape, save deserts and tropical forests. They have even expanded into human society, where they endure in an altered, domesticated form.

Sensing the intruder's approach, Mongo breaks ranks and trots down the ice staircase as the men gaze on in wonder. Dog (*Canis lupus familiaris*) and wolf (*Canis lupus arctos*) greet one another, then happily wander off together, past the drifts and hummocks, leaving the men with little hope they will ever see their pet again.

Unlike foxes, wolves have been extinct in the British Isles since about 1700, where popular knowledge of these animals has been reduced to Grimm's fairy tales and Aesop's fable "The Boy Who Cried Wolf." The Bibles aboard the

Investigator contain numerous references to wolves, which are often cast as metaphors for destructiveness and greed. Given the chance, wolves will attack and kill livestock. This, and a propensity for carrying rabies, is the foundation for the men's native hatred and fear.

Soon, the pair are seen making their way back to the ship, stopping time and again to play. The men keep the wolf in their sights, but the animals remain so close together it is impossible to risk a shot. Armed with boarding pikes, the crew stand ready to rush down and save the dog. But then the wolf disappears, and Mongo lopes back to the Investigators, tail awag. He is collared, tethered, and then sent out as bait. Over the next few days, the wolf returns to play on several occasions—and then quickly disappears among the shadows and white.

The time has come to put idled manpower to use in the search for Franklin and the further exploration of the region. McClure gives the order to begin preparations for three ambitious sledge journeys. The work begins with the establishment of a depot on shore for the sledge parties should they find themselves cut off from the ship. The cache contains a whaleboat and provisions for sixty men for a period of ninety days. Because it also contains a considerable stock of rum, the entire island is placed off-limits to prevent sailors from raiding the stash.

As they await their marching orders, the men spend their time on the hunt. These excursions away from the vessel are deemed good for both the circulation and the soul. While the officers are more or less competent hunters, the seamen are not. As members of Britain's working classes, they have almost no experience with this gentleman's game. In Miertsching's assessment, the sailors are—with few exceptions—virtually inept. It is a blessing they are not relying on their skill for survival. They track, but have little success shooting ptarmigan, hare, and fox. When game is found, the brother often distinguishes himself with his marksmanship. Often, he walks slightly apart from the others or all alone, thinking of the brotherhood in Europe, or of Esquimaux friends in Labrador hunting under the shared, limitless sky.

Miertsching returns from the hunt one day to learn that Armstrong was "properly humiliated" by his fellow officers. No one seems willing to name the exact nature of the offence against him, but its seriousness can be read on the faces of the men. Although the brother deplores rough treatment of any kind, he believes that it was not done without cause. The surgeon is not welcomed company and has trouble keeping his opinions to himself. It seems that he finally went too far. When McClure is informed, he listens without expression, unwilling to act or acknowledge. The officers swiftly close ranks and conduct themselves as if nothing occurred.

It isn't long before the officers find themselves in need of the surgeon. They stumble down to sickbay with burning, teary eyes. Some don't notice symptoms for several hours following their return to the ship. Then the redness, itching, headaches, and pain set in. It is as if their eyes have been lashed with sand. Some complain of hazy vision, others see halos around sources of light. A few lose their sight all together. The hunters are suffering from *niphablepsia*—snow-blindness—caused by the intense ultraviolet rays reflecting over the polar ice and snow. Up to 90 percent of the sun's rays that strike the cryosphere are reflected out and away, preventing the land and sea from absorbing the solar energy. Along the way, it strikes and burns the men's corneas as they scan for prey. Armstrong calms their fears of permanent blindness, ordering them to remain in bed with patches over their eyes. Symptoms lessen over a day or two, leaving them profoundly grateful when the stinging stops and their vision eventually returns.

Long ago, the Inuit developed an effective defence against sunlight reflected off ice and snow. They wear goggles carved from caribou antler. A thin, horizontal slit is cut through the shield, permitting only a small band of light to pass. This shield is held snugly in place by a cord made of caribou tendon tied around the back of the head. Onboard the *Investigator*, only a few of the officers have dark glass goggles. Armstrong issues medical gauze to the entire ship's company. He orders the men to wear it as a veil while out hunting in the light of day.

On March 31, the captain and surgeon return from a hunting trip to find the *Investigator* disrobed, exposed, the deep snow wall and resulting drift removed and hauled away. The ship now looms as a determined and defined silhouette against the surrounding sea of white. The following day, work parties remove the ice from the upper deck, revealing the planks once more. The air is filled with the tang of bubbling pitch and the crack of hammers as the crew recaulks the seams. This task, and the imminent departure of the sledge parties, has the entire company in motion.

On Good Friday, divine service is held in the morning. The ice is measured at six feet, five inches thick. Although the temperature peaks at 28 degrees, there is no doubt that there is still more winter to come. As last-minute preparations are made, a solemn mood settles on the crew. Many men believe that this will be a final parting. The sledge crews fear they will return only to find that the *Investigator* has been swept away, shattered in the ice, or disappeared to the bottom of the sea. The entire ship's company takes its midday meal together, as a family, then assembles on deck for inspection and encouraging words. McClure distributes the gift of a pocketknife to every man onboard. He orders the ensign raised as the sailors strap into their harnesses.

Ten feet long, one foot high, and thirty inches wide, the sledges weigh sixty-four pounds apiece. Onto each is loaded food, fuel, and clothing for forty days—a weight of 1,028 pounds. The commanding officer of each party will walk in front of his team of six men, seeking the easiest route through the maze of rubble and ridges. He will carry a double-barrelled gun, telescope, compass, and notebook for recording observations. First Lieutenant William Haswell is ordered to travel southeast along the coast of Prince Albert's Land. Ship's Mate Robert Wynniatt is to proceed northeast to search the southern shores of the strait. Second Lieutenant Samuel Creswell will head to the northwest coast of Banks Island.

Armstrong wonders why a team has not been dispatched north to Melville Island, little over a hundred miles away. Captain Austin and his four ships left England under orders to reach it—why isn't McClure attempting to make contact? More importantly, if Franklin had been forced to abandon his ships in this uncharted region, wouldn't he attempt to reach the location where Parry famously wintered? The surgeon carefully registers these concerns in his journal.

The sledge crews are sent off with three rousing cheers. A fatigue party aids each for several miles until at last the men stop, exchange harnesses and handshakes, then go their separate ways.

Eighteen days after setting off, Wynniatt's sledge returns at one o'clock in the morning. The Investigators rise to greet him. His men are in good condition and they were making reasonable progress. Then, about a hundred miles out, the lieutenant fell from a hummock and broke his pocket watch. Unable to make accurate reckonings, he thought it best to return to the ship. McClure listens to this excuse in silence, then unleashes his fury. Why didn't he continue along the shore as ordered? His primary purpose, after all, was not to chart the coast but to search for Franklin. He had hoped his troubles with this insolent officer were behind him. No, McClure declares, there are no spare pocket watches onboard the *Investigator*.

"Before chronometers were in use, men were able to find their way all over the world and discovered new lands."

The captain continues to berate his young lieutenant, who withers in shame. Wynniatt and his crew will be permitted twelve hours' rest, then ordered back out on the ice to resume their trek. The brother witnesses the entire exchange, but fails to offer his watch.

In a sign of the shifting season, a dry spell allows for bedding to be hauled up on deck and aired. Soon, a lemming is found with its coat a blend of snow white and tundra brown. Arctic foxes appear like smudges on the icescape, similarly caught between winter and summer hues. The surgeon's autopsies of several reveal their stomachs mostly distended and empty, apart from several pieces of Arctic willow in one and bits of caribou hoof in another. The appearance of the first snow bunting is widely considered a harbinger of spring.

A seal pops his head up the fire hole, only to be speared by a boarding pike. Upon examining the body, Armstrong finds the remains of numerous old wounds on its hide, possibly the result of sparring with rivals or a narrow escape from a polar bear. The meat is consumed, the blubber given to the dog, the skin and skeleton cleaned and preserved for transport back to England.

Polar bears are seen with increasing frequency. Soon, one is spotted sauntering directly toward the *Investigator*. The officers are out playing a game of cricket when someone sounds the alarm. They grab their jackets and make a dash for the ship. Once everyone is safely onboard, hunters are dispatched to hide on the ice. Marksmen are posted on the stern. As the target steadily advances, he pauses now and then to raise his snout and inhale the foreign odours. When he reaches within a hundred yards of the ship, McClure fires a ball into his back. The beast falls, but finds his legs again. He writhes violently, then sits on his haunches as the men on the ice close in. At last, Armstrong shoulders his rifle and sends a ball through the animal's brain.

Men crowd around, admiring the bear's coat, the colour of clotted cream. They run fingers through the glossy, translucent guard hairs and dense under-fur. The coat sheds water easily, allowing its owner to shake like a dog, rapidly expelling gallons of water out and away from the skin. Behind black lips are large canine teeth and serrated molars, the dentition of a confirmed carnivore—the largest that roams the Earth's surface. The men compare the paws and claws with their own hands and fingers. Polar bears are powerful swimmers, using their huge, dish-shaped forepaws to paddle well over a hundred miles between floes. Small bumps and hollows on their paw pads act like suction cups, providing superior traction as they swagger across the ice with the quiet, confident mastery of all they see.

The bear is measured and weighed: an adult male of seven feet, nine inches, seven hundred pounds. Born at less than two pounds, polar bears can grow to well over fifteen hundred. Polar bears range throughout the circumpolar world, from near the geographic North Pole south to Newfoundland and Labrador. Occasionally, they are even spotted in the Gulf of St. Lawrence. Polar bears are perfectly adapted to—and completely reliant upon—Arctic ice for hunting ringed seals, their primary prey. They will wait patiently at holes in the ice, often dispatching adult seals with a single, fatal snap of the paw when they surface to breathe. Seal pups are found by crushing through maternity lairs, which they can smell from twenty miles away. In the absence of food, polar bears are able to significantly slow their metabolism until the next meal is found. They have only hunger and humans to fear.

The Inuit hunt polar bears for food and as a young man's rite of passage. The ability to face and conquer such a formidable opponent—armed only with a spear—speaks of a bravery that transcends cultures and time. It also builds the kind of confidence necessary to go out and provide for one's community under the most extreme and unforgiving conditions.

The Investigators lose no time removing the bear's skin, revealing a layer of white-yellow fat underneath. Stripped, the muscular and skeletal structure looks uncomfortably familiar. After the pelt is pulled free, the surgeon orders the stomach split open. He squats down and fingers through a few steaming raisins, cubes of pork fat, tobacco leaves, and pieces of adhesive medical plaster. The cook is called to give his professional opinion. He declares that the pork fat appears to be the remains of mock turtle soup, common naval fare. The items all seem to have been recently consumed. Presented with this discovery, McClure sees two possibilities: either the bear encountered rubbish jettisoned from the *Investigator* last autumn, or the *Enterprise* is wintering nearby. Intrigued with this second hypothesis, the captain orders Cresswell and his party—minus the invalids—to repack their sledge and explore along the southeast shore of Banks Island.

Two days following Cresswell's departure, hunters discover a large tin can a few miles across the ice. Clearly, a castoff from the *Investigator.* The surrounding snow is covered in tracks and evidence: tiny scraps of food and refuse corresponding to the polar bear's final meal.

On May 20, a party is observed approaching from the north at 7 a.m. This group is not due back for another week, and its approach is cause for concern. Cresswell returns with two casualties aboard his sledge. The sailors persisted in their harnesses far past the point of sense and duty. They were finally ordered to ride. They now suffer from severe frostbite and are no longer able to walk. Both are feverish, and their throats are inflamed from eating ice and snow. Despite his best efforts, Armstrong is forced to amputate several fingers and toes. It is only with luck and constant care that the rest of their hands and feet are spared.

Nine days later, First Lieutenant Haswell is spotted at eleven o'clock in the morning. Over a period of forty-one days, his party made a roundtrip of 720 miles southeast along the unknown coast of Wollaston Land. To the surgeon, they appear in better health than when they left, having had the benefit of abundant driftwood and game. The men explored the remote coastline, travelling into numerous bays and inlets. No trace of Franklin was found.

They did, however, make one notable discovery on their return. About six days southeast of the ship, they came across an Esquimaux sealing camp. The encounter with these bewildered people was a friendly game of charades. Strangely, the Esquimaux did not seem to understand the custom of nose-rubbing. Unable to glean any information, the Investigators traded buttons for seal pelts and then continued on their way.

Upon hearing this news, McClure turns to Miertsching. The brother requires no encouragement. A sledge, six men, and rations for twelve days are made ready. They leave at 6 p.m.

At midnight, McClure calls a halt to the march. In the still twilight, all perspective is lost. The distinction between pastel sky and ice is diminished. They exist at the centre of an all-but-seamless void. After a meal of bacon, grog, and a few pulls on the pipe, the sailors slip back into their harnesses. They lean forward and follow their captain and his interpreter across the frozen sea.

10.

THE RED SCARF

With temperature spikes to 45 degrees, the land is emerging from winter. Plentiful driftwood and consistent good weather allow McClure and his men steady progress along the shore. Five days out, the only cause for concern is the captain's coxswain, who is having trouble with his feet. He marches through the ice and snow in thin canvas shoes. And then Miertsching sights what appear to be tents on a hill overlooking the sea. He can hardly contain his joy. The Investigators are surprised to find Esquimaux here—those met by Haswell were six miles off shore. The captain orders the men to alter course and follow the brother's lead.

As they approach the Esquimaux camp, Miertsching feels the weight of responsibility descend. Will these people understand the words he's brought all the way from Labrador? Five men, five women, and a number of children can be seen. They stand in a cluster near their sealskin tents, gaping as the intruders close in.

"We are visiting you as friends," the brother declares. "We bring rare gifts."

The Esquimaux are immobile, in evident shock. They offer no hint of comprehension.

"We are afraid," they reply at last.

The brother smiles. These words he knows well.

The Esquimaux are unarmed and show no sign of resistance. As he closes the gap, the brother pays out his most calming introductory lines. Honeyed words of greeting, friendship, peace. Surely, they will detect his sincerity. This proclamation is followed by the display of gifts. Soon, he stands among them. They surround the stranger, inspecting his hands, hair, face, the cut of his sealskin coat. At last, he is accepted as fully human.

The Esquimaux explain that some of their men are out hunting on the ice. Others are on the land, chasing herds of muskoxen and caribou, which are plentiful but difficult to approach. They say they live in snow houses during winter; in summer, they shelter in these sealskin tents that are easily moved from one place to the next. High-quality copper is obtained in trade from Esquimaux to the east. This is worked into harpoons, arrowheads, hatchets, needles, and knives.

Their features, language, and tools are so very familiar. Even the women's parkas, with their long tails, match those found in Labrador. The women previously met on the mainland coast wore shorter and broader tails in the style of the Greenland Esquimaux. Although these folks do not wear labrets, the women adorn their faces with thin blue and red tattoos. They have never met a stranger and take Miertsching's mute, oddly attired companions for supernatural beings. The brother promptly corrects this impression. Still, they are astonished to learn that there are other inhabited lands. They supposed they were alone in the world. On clear days, they can see a great land to the south, but it remains a mystery.

A looking glass is produced and passed from hand to hand. Astonishment and delight brightens each face as the people take turns admiring their own reflections. One of the sailors demonstrates the use of gunpowder, firing his musket out over the sea. The resulting fright takes time to settle and calm.

Having at last regained their confidence, having answered their many questions, the brother borrows an enormous sealskin and lays it out flat on the ground. Over this he spreads a large sheet of paper and begins to make a sketch. Esquimaux and Investigators gather round. He draws an outline of the coast they have just travelled, along with the position of the ship and the location of their camp. He points to features on shore and their corresponding representation on the map. After a great deal of explanation, he believes his hosts grasp the essence of cartography. He hands the pencil to one of the men, who then reaches out to touch the paper.

"What sort of skin is this?"

It is the same question they have about the Investigators' woollen clothes, canvas tent, and cotton handkerchiefs. Textiles, agriculture—where to begin? The brother shifts the subject.

McClure and the rest of his men understand none of the language burbling around them. They only know that this must be the first instance of paper and pencil being used on this shore. Do these simple people have the slightest clue what Miertsching is trying to explain, or are they humouring him? McClure grows impatient and sets off to explore the surrounding hills.

The brother upends his entire vocabulary, picking through words he may have missed. *Simply draw a line that continues the coast to the limit of your available knowledge.* Like the sailors in school, these pupils grasp the pencil awkwardly but are keen to give it a go. Both men and women offer correction and advice. After an hour, the map is complete. In the opinion of everyone involved, the result is both pleasing and true.

When the brother considers what they have done, he feels vindicated. Together, they have bridged a gap in knowledge. The shoreline they have sketched connects all they way to areas previously explored: Point Parry on Victoria Land, as well as Sutton and Liston islands in Dolphin and Union Strait. These are all marked correctly. Referring to the map, they recite the names of capes and other significant features. They point to the line depicting the southeast shore and tell of Esquimaux living there. A list of names is offered. The result has exceeded expectations. Through the art of communication, the brother has proven his value to this expedition. And now, while he awaits the captain's return, he turns to his other, greater duty.

News of a universal creator has not reached these people. Their beliefs correspond to those of the Esquimaux met on the mainland shore and in Greenland. After death, there is both a good land and bad land where people go according to their earthly conduct. Meanwhile, the sun looks down from the nighttime sky through the holes of the stars. Added to these known beliefs is a new story, passed down from ancestors, that tells of a high mountain where the people went to live during the time of a great flood. Miertsching takes this as evidence that even here, on the furthest reaches of human existence, a biblical event is remembered and confirmed.

McClure returns and is impressed with the Esquimaux map but impatient to be underway. The brother—hoping to relay the story he holds so dear—pleads with the captain for a few more hours. Such a revelation takes time. The weather is clear, and it is impossible to know when such an opportunity will present itself again.

Under other circumstances, McClure would readily agree. But the coxswain's condition is deteriorating. He had to be carried in and out of the tent this morning and now sits on the sledge in his stockings. Keenly aware of the damage done to Cresswell's men, McClure is unwilling to delay. The brother reluctantly accepts the order. Once again, someone else will have to do the Lord's work. Someone else will have to make the incredible journey to reach this place, find these wandering people, speak their language, earn their trust, and then instruct them about their creator. Someone else will have to do all this before men of other ambitions arrive to take what advantage they can.

The brother turns and distributes gifts. Knives, saws, needles, beads, red and blue flannel. He leaves the looking glass, which will no doubt provide hours of entertainment to come. With each gift given, the recipient wants to know its value and offers to part with his best copper implements in return. They do not understand that these are gifts given without expectation of payment. McClure finds this humbling in the extreme.

The Esquimaux follow them down to the floe, watching the strangers strap themselves to the sledge. As the Investigators thank them for their hospitality, McClure is overcome with emotion. The men are surprised by his tears. To McClure, these people appear hopelessly impoverished, lost at the frozen end of the Earth. The hurried visit from civilization, the trifles given, it all seems so horribly inadequate. He believes it is his duty to help them, only he is unable to do it.

Neither he nor Miertsching consider the possibility that they themselves may be the objects of pity. The Esquimaux stare back at men used as dogs to haul a heavy sledge across the melting snow. One of them has clearly suffered from his exposure to the elements in thin, inadequate clothing. He is now a burden to the rest of the group. And where are they going? These men travel the ice, away from abundant game and seals, without women to help and care for them, without children to support them in their old age. How long could such a people last?

McClure stops and removes his thick red scarf. He wraps it around the neck of a young woman with a baby on her back. This will provide a modicum of warmth. It is the least that he can do. The gesture startles the young mother, who declares that she has nothing to give in return. Seized with grief, she reaches behind and pulls out her baby from her hood. She covers it with tears and kisses, then offers it to the captain in exchange. McClure steps back, confused.

"No, my lady!" the brother exclaims. "Please understand. This is a gift from the captain to you. Nothing is expected in return. Your child is yours to keep."

At this she brightens, returns the baby to her hood, and

considers the captain anew. She laughs, then fingers the woollen scarf. Finally, she says something in her language. McClure looks to his interpreter.

"She asks, 'What sort of animal has red skin?'"

At this, the captain smiles, turns, and continues on his way.

11.
WATER SKY

Miertsching lifts the telescope and focuses across the ice. These past few weeks, he has lost faith in his powers of observation—his own eyes are not to be trusted. But in this case, the image seems plainly clear: a polar bear reclining a few hundred yards from the deck of the *Investigator*. Through the series of finely ground lenses, he can see the animal's face, its black eyes and nose. It frequently moves its head from side to side as if to sniff the air. The brother passes the telescope to the men nearby, and they each come to the same conclusion. Rifles are loaded, and soon a hunting party is splashing across the rotting floe. Then, when they approach to within 150 paces, something miraculous occurs: their quarry suddenly rises into the air and flies off, leaving the men soaking in knee-deep water as a snowy owl ascends into clear blue sky.

They have been fooled yet again. Fine weather and temperatures soaring into the forties have brought an increase in the appearance of remarkable and convincing mirages.

The air just above the surface of the ice is much cooler than the ambient temperature, and the light travelling through these layers has a tendency to bend. Observing objects over the surface of ice in these conditions can make them appear much taller than they actually are. It can transform distant pressure ridges into an archipelago levitating above the horizon. Far-off hummocks appear to rise into turrets and spires. If approached, the resulting ice castles remain forever out of reach. These illusions are known as *fata morgana,* Italian for "Morgan le Fay" (Fairy Morgan), the sorceress half-sister of King Arthur, who was thought to dwell in just such a place. The effect can even change birds into bears.

It is mid-June, and for the past two weeks the ice has been covered with pools and streams. The cryosphere is in retreat. Only a week ago, the men were able to march to shore with ease, allowing them to explore the moss-covered plain and discover the ruins of thirty-two ancient stone dwellings, along with stone lamps, arrowheads, and blades—artifacts of a once-abundant people. Ten miles inland, an ancient sledge was found with runners made of bone.

The *Investigator*'s various sledge parties have all returned. On the journey to meet the Esquimaux, the captain caught a severe cold and remains confined to bed. Two members of his sledge team join the casualties of Cresswell's expedition in sickbay, recovering from frostbite and the loss of fingers and toes. Additional losses resulting from these journeys include provisions nicked by foxes, the surrender of boots and clothing to a hungry bear, and the destruction of Wynniatt's chronometer. All told, nearly five hundred nautical miles were explored, and not a single trace of Franklin was found.

The men are buoyed by the hope that they will soon be on their way through the Passage, and home to England before the year is through. While they await their release, they channel their energies toward more fanciful pursuits. Half the crew can be seen at a time out flying brightly coloured kites under the midnight sun. Others build small boats of wood and cork to sail in the growing lakes. These friendly regattas come to an end one evening when a sailor turns pirate and successfully attacks and destroys the entire fleet. He is enthusiastically pursued for an hour across the wet and slippery floe.

The crow's nest is hoisted a few days past summer solstice. A stretch of strong wind and frost temporarily halts the melt, but good weather soon returns. The captain orders that a new jacket, trousers, underwear, jersey frock (sweater), socks, and boots be given to each member of the ship's company. Everyone seems pleased with the gifts—a new wardrobe in which to greet their emancipation.

Over the past month, the thickness of the ice has decreased 26 ½ inches. The snow continues to melt on shore until only a few white shadows remain in protected hollows and ravines. Cracks soon appear near the ship, some large enough to swallow a man. The changing pressure shears a nine-pound sheet of copper from the hull.

The first open water is observed along the eastern shore on July 7. This sighting coincides with the appearance of the season's first mosquitoes. The crew welcomes these pests from the old, familiar world. Their arrival is yet another sign to bolster faith. Hoping for reassurance of a more substantial kind, Assistant Surgeon Henry Piers seeks out the opinion of the oldest and most experienced member of the crew.

"Get home, sir?" the quartermaster replies. "To be sure, we will. Because if we don't get through this time, we never shall."

The break is so gentle it takes them by surprise. At 2:40 p.m. on July 14, the ship disengages from the floe to which she has been attached for the past ten months. The entire process takes ninety seconds. Some of the men have gone to collect their clothes, washed and left to air out on the ice. Armstrong is able to scramble aboard just before final separation occurs. Some men do not make it in time and must be hoisted up from the edge. Numerous items are abandoned in the great rush toward the now-moving vessel. Soon, however, all hands are onboard and accounted for.

As the *Investigator* lurches and settles into the rift, she is joined by a herd of seals exuberantly chasing each other through the newly opened water, up and over moving chunks of ice, and back into the fluid sea. It is as if they too rejoice at their sudden release.

For the next three days, the *Investigator* drifts with the current and tides within five miles of her winter position. On July 17, the call "all hands make sail" is heard for the first time this year. Five days on, the rudder is shipped, but they are still unable to sail. For protection against the onslaught of shifting ice, they anchor to a large floe—which promptly runs aground and splits in two. Trapped at the edge of the ice stream, their situation is precarious, but soon both ship and ice begin drifting in the desired direction: north toward Melville Island and the completion of the newfound Northwest Passage.

To free the ship entirely, a hole is bored in the adjacent floe, and a cask packed with forty-seven pounds of powder is placed inside. A long fuse is laid out and lit. Eleven minutes later, the explosion sends shards of ice and a saltwater geyser a hundred feet in the air. The shock rings bells onboard. It results in a hole twenty-five feet across and a series of cracks radiating two hundred feet beyond.

Despite this dramatic assault, the ice soon closes in again. The rudder, in danger of being damaged, is unshipped and brought aboard. The floes crowd and crash their competitors, pushing their way atop one another while smaller bits are shoved under and down. While the rumbling is ominous, it does not reach the fury heard at the end of last season. This time, the men tell themselves, they will not be so easily frightened. But then the compass grows sluggish and fails, the mist moves in, and it becomes impossible to tell which way they are going. When the fog finally lifts, a pair of muskoxen are revealed contentedly grazing the near shore.

On August 2, the anniversary of their first encounter with the ice, the strait to the north remains packed as far as the eye can see. At dinner that night in the officer's mess, the toast is to "more water." This call is soon answered with rain, but no hint of water sky—that welcomed bruise that appears on the underside of low-lying clouds, indicating open, navigable seas below. It is the converse of iceblink, the distant, dazzling band on the bellies of clouds basking in light reflected off ice and snow. And now the season of light itself is on the wane. The sun sets behind the western hills at half past ten. This realization stirs the collective unease.

Knowing it is impossible to blast their way north through the strait doesn't stop well-armed men from trying. On August 14, a cask with thirty-six pounds of powder is placed at the centre of the anchor floe and produces the most satisfying blast thus far. It allows the ship room to manoeuvre and the rudder is shipped once more.

Progress north, however, proves impossible. After two more days in the attempt, McClure announces a change of course. He now sets in motion the most fateful decision of his career. The order is given to bring the ship about. They will make Melville Island by sailing south, then up the west coast of Banks Island, ultimately approaching from the west. The *Investigator* is turned and her sails unfurled. Soon, the ship is travelling at six knots—fast enough that water washes back over the foredeck. Now the route is nearly ice-free, and the Princess Royal Islands—points of reference and constant companions throughout the long winter—can no longer be seen.

The exhilaration of sailing emboldens the crew. Travelling close to shore, progress is swift, and hope rises to the surface again. Polar bears are spotted swimming among the loose ice, along with the spout and fluke of whales. On shore, mortal combat between a fox and a goose can be seen, with the goose unleashing a dramatic assault—most likely in defence of goslings.

Some of the geese have pure white plumage and wingtips that appear to have been dipped in ink; others tend toward dust grey-blue. The men observe them travelling together in large flocks, issuing loud, nasal honks in chorus, forming undulating lines or a giant U as opposed to the pointed chevrons of other waterfowl. Snow geese have made their cacophonous migration to the Arctic from as far south as the Gulf of Mexico. Here, they congregate in large colonies, breed, and lay eggs atop nests fashioned from moss, willow, grasses, and down. A mating pair will vigorously defend its clutch from competitors on the lookout for a ready-made nest and predators on the hunt for an easy meal. The Investigators continue to watch the contest until the goose succeeds in her valiant defence and the chastened fox turns tail and flees.

The south shore of Banks Island is soon rounded and the *Investigator* is again pointed north and east. To port, the unending polar pack; to starboard, gorges and hills caught up in the full flush of the Arctic summer. The flowers and grass attract caribou, a diversion that helps take the men's minds off their predicament. If the winds shift and the vast continent of ice encroaches on the land, the only thing standing in its way is the *Investigator*.

The Polar Sea is an unending terrain of rolling hills and pressure ridges rising as much as a hundred feet above the surface. Even the near ice reaches as high as the ship's lower yards. Should the floes rush in, the best that can be hoped for is that the ship will be thrown broadside, up on the beach, where the men will be forced to winter in this fully exposed position. Some speculate that such a fate could have befallen the *Erebus* and the *Terror*, although no wreck, cairn, pole, or the slightest evidence of any previous ship is found.

The keel often strikes sandy bottom, and hours are spent wresting the ship away. On one stoppage, a party of officers and men are sent ashore to explore. They scatter over the beach and hills, leaving Assistant Surgeon Piers and Ice Master Court together, enjoying long strides over the grass, lichen, and Arctic willow. They pause now and then to crush and inhale the scent of these fresh, growing things, marvelling at the acidic-sweet taste on their slumbering tongues and palates.

First Mate Hubert Sainsbury rushes up to show them a collection of petrified wood he has found nearby. He is soon followed by a sailor with a pair of haversacks and orders from the captain to collect as many specimens as possible.

Near the top of a hill, the men discover the upright trunks of numerous trees and logs lying embedded in the earth. One trunk measures five feet in circumference. Some of the pieces are as heavy and hard as stone; others are light and brittle, and appear to be mummified. In a nearby gorge, more wood is found. Surveying the scene, the assistant surgeon wonders: could trees once have grown on the edge of the Polar Sea?

The men have wandered into the remains of a forest of great antiquity. Stumps, logs, seed cones, and leaves have all been preserved where they grew some fifteen million years ago. Pine trees once dominated this landscape, but spruce, larch, and a kind of cypress were also present. The average height of the forest was perhaps seventy feet, and its composition would have been comparable to modern, temperate forests currently found a thousand miles to the south along the coast. Recognizably modern mammals—including wolves, raccoons, horses, deer, and camels—possibly passed between these trunks; owls and crows likely perched in the branches above. The presence of these trees points to a climate that was significantly wetter and as much as 40 degrees warmer than the current mean average. The forest most likely died in a climatic shift and wave of extinctions that occurred some 14 million years ago. The event—the Middle Miocene disruption—was perhaps triggered by a meteor impact in Europe or a volcanic eruption in Africa. Whatever the cause, the Arctic climate underwent rapid and fundamental change, leaving an utterly transformed landscape—and these remarkably well-preserved remains.

Piers, Court, and Sainsbury fill their pockets and haversacks with specimens and begin the hike back to the ship. The distant boom of the signal gun encourages them to quicken their pace.

Onboard the *Investigator*, they lay out the evidence of their curious discovery as their fellow officers gather round. To Armstrong, the ship's ranking scientist, it seems improbable that these trees could have floated into their current position. He suspects that the Arctic climate must have been significantly warmer in previous times, but it is a mystery for some future scientist to solve. Before he or anyone else has a chance to voice his theories, the ice suddenly shifts—hoisting the ship six feet out of the water.

Blasting will not alter this situation. There is nothing to do but wait for the ice to relax and release.

The following day, the men are ordered ashore to collect the finest specimens of petrified wood for the Botanical Museum, as well as a stockpile of mummified wood for the stove. It is discovered that logs buried a little deeper in the sand and clay resemble brown coal. When burned, they give off a similar odour. Great quantities of this wood are hauled aboard and stowed. It may be needed to keep them warm over the coming winter.

The coming winter. The coming *polar* winter. It is a thought that each man tries to push down, to deny. But the cooling weather weakens their resolve. At night, young ice now forms in the pools on shore. Observing the signs, the brother momentarily succumbs to a darkening mood.

Should this place become our winter quarters, the ship will likely be our tomb.

August 29 dawns with strong northwesterlies and gusts of sleet and snow. The pack is on the move. At eight o'clock in the morning, heavy pressure bears down on the anchor floe, spinning it completely around from its grounded position, projecting it twenty-five feet above the surface of the sea. Ice now overshadows the deck of the *Investigator*. As the crew rushes up to witness the spectacle, the floe suddenly splits in two. Following a pair of dramatic rolls, the half to which the ship remains attached drags them into deeper water amid heavy, moving ice.

All hands are required to keep the *Investigator* secured to the floe and to unship the rudder, which has been seriously damaged. A line is readied on the main yard for the evacuation of provisions. The men are assigned to boats and prepare to abandon ship at a moment's notice. Throughout the afternoon, adjacent floes are attacked with pickaxes and chisels. When this fails, powder is deployed.

At last, the pressure suddenly shifts and the ship drops two feet. Assistant Surgeon Piers is among those working on the ice when this occurs. He makes a run for the deck that now lists 15 degrees. The *Investigator* is then driven into the anchor floe and begins to complain as never before. The mizzenmast shudders overhead. Planks of teak groan and vibrate beneath their feet. Startled bells ring. Armstrong grabs hold of the capstan and feels a tremendous wave of gratitude for the men who designed and built this sturdy ship.

Lying in sickbay below, Boatswain George Kennedy has been nursing a nasty powder burn across his abdomen. The accident occurred a few days ago while he was loading his gun on shore. Convinced he had been mortally wounded, he still wonders how he survived. Now, it seems as if his luck has finally run out. The crack and roar of unseen ice is too much to bear. At any moment, the leading edge of a floe will surely breach the hull and crush him against the bulkhead. He swings his feet over the edge of the berth, gingerly pulls on a shirt, and staggers up on the pitching deck.

The boatswain is followed by the captain's steward. No longer willing to wait for the aft cabins to be crushed, he hastily packs the captain's carpetbag and makes a run for the ladder.

The carpenter now steps off the deck and onto the floe to survey the damage. Copper sheeting hangs in shreds in some places; in others it is neatly rolled up like a paper scroll. The ship is then knocked hard about with her bow facing the shore. The crowding and crashing floes run their huge bulk into and over their competitors, driving them under the bow of the ship, further increasing her list to port.

"This is the end," McClure declares. "Our fate is sealed. The ship is breaking up—in five minutes she will be sunk!"

The brother sits with his back against the gunwale of the listing ship. He looks to the men sitting beside him—they stare back with wide eyes and pale faces, shaking.

With a trembling voice, the captain gives the order to cut the five hawsers holding the vessel to the ice. His last hope is that the wreck might be thrown sufficiently high and clear to at least serve as shelter.

As if in response, the ice begins to quell. For a few

"First discovery of land by HMS *Investigator*. September 6, 1850." This drawing, like the one below and on the following page, was made by the ship's second lieutenant, Samuel Cresswell.

"Critical position of HMS *Investigator* on the North Coast of Baring [Banks] Island, 1851." The men of the *Investigator* knew that they could remain stranded like this for at least eleven months, perhaps the rest of their lives.

In Mercy Bay on April 15, 1853, a sledge-party led by Creswell departs from HMS *Investigator*. When large pressure ridges and heaves had to be crossed, the team was reduced to crawling on hands and knees.

On two occasions, the head of a polar bear appeared inside a tent and sniffed at the sleeping men packed tight in their blanket bags.

The flesh and blood of caribou helped sustain the men of the *Investigator* through years of desperate need.

The men had no love for the gyrfalcon, and shot one that threatened a flock of small songbirds.

The "Investigators" kept an Arctic fox as a pet. Old Adam, as he was known, became so tame that he allowed anyone to touch him. Other Arctic fox were skinned and consumed.

Robert McClure, captain, HMS *Investigator*: "See how holy writ mocks me. In this crisis and extremity, when all our lives are trembling in the balance, I opened the Bible to find words of comfort, and this is how it answers me..."

'ohann Miertsching, a missionary of the Moravian Brotherhood, was chosen for the expedition ıot as a spiritual guide, but because of the five years he had spent among the Inuit of Labrador ınd his mastery of their language.

Renderings of satellite images of Arctic sea ice from September 1979—the first year that reliable images were available—and, below, from September 2007, showing an open Northwest Passage.

moments, the truce seems to hold. Then the floe under the stern cracks, the shock of which sends the carpenter lunging back onboard. The *Investigator,* still firmly gripped, is now slowly carried toward a great heap of ice on shore.

"Maintain your posts!" McClure shouts before retreating to his cabin, leaving the men to ponder their fate on their own.

The sea remains calm at dusk. The men are ordered to their hammocks below, but are to remain fully dressed. Half an hour later, the pressure returns and they are called back on deck. When an approaching floe is deemed no longer a threat, they return to bed and fall into a fitful sleep.

They rouse in the stillness of dawn. Three fathoms of water have formed under the bow, but the ship remains trapped a hundred yards from shore. Soon, all hands are engaged against the ice. Armed with pickaxes and shovels, men roam the battlefield through falling snow like a seasoned team of sappers. After a target is chosen, they gather to hack and dig—then hustle away before a series of choreographed explosions sends boulders and flakes rocketing into the sky. A hundred pounds of powder are expended in the fight.

Miertsching and Piers walk to the end of the floe. There, they consider the catastrophe of ice at this edge of the Polar Sea. It is a scene neither man expects to forget. It reminds them of a town levelled by an earthquake. Here, the wreckage of bricks, columns, and slate is replaced with blocks and slabs of ice three and five feet thick. The fresh rubble piles reach ten feet above the waterline. Past this immediate destruction zone, a mile-long floe runs parallel to the ship. In the pack beyond are a series of hummocks and ridges—the limit of which is any man's guess. They suspect they are the first Europeans to take in this apocalyptic view. What remains unclear is whether either of them will survive it.

An extraordinary hush descends. No one speaks more than necessary. The crew seeks to relieve its mounting frustration through another day of blasting. This work endows the men with a sense of power and possibility. It gives them something to do.

Miertsching feels compelled to counsel McClure. Despite the brother's own doubts and fears, despite the expected resistance or lecture on the unsuitability of land-based Christianity, he prepares himself to offer up some hopeful message. The captain saves him the effort.

"I have arrived at the genuine conviction that a Higher Power, an Almighty Providence, is watching over and shielding us," McClure declares, "and that under this protection, none of us will lose our lives. We will all reach our home and families again. Of this, I am certain."

By September 1, what little water had been observed between floes has frozen over. The mist and cold present a gloom to match the morale onboard. Dreams of sailing for home this year have fallen away. Progress in either direction now seems impossible. Even the prospect of retreating to Barrow Strait next summer seems remote. All told, they managed just four days of open sailing this year. Regardless, McClure steels himself and then commits to his men that he will do everything in his power to make their lives comfortable and happy until their eventual release.

A flock of snow buntings appears to be gathering provisions for the journey south. These familiar white and black songbirds forage for both insects and seeds. They nest in rock cavities throughout the Arctic, as far north as latitude 83 degrees. In winter, they spread south into North America and northern Eurasia. A lucky few even reach the British Isles.

The men are not alone in noticing the buntings' arrival. A gyrfalcon hops off its rocky perch and flies into the scattering flock. A member of the world's largest falcon species, this gyrfalcon preys primarily on ptarmigan, although most any bird will do. In its pursuit, it seems to defy the rules of aerodynamics under which most birds are constrained. On pointed wings spanning four feet, it performs spectacular aerial displays with steep dives and rolls of 180 degrees—a flash of white breast and mottled back, the constant and formidable gaze. Once the exclusive property of emperors and kings, gyrfalcons have long been prized for falconry. This bird—a year-round resident of the Banks Island shore—is likely responsible for three or four hungry chicks waiting on the cliff nearby. As the other men watch its relentless assault on the smaller, beloved birds, Wynniatt raises his rifle, pulls the trigger, and drops the falcon from the sky.

Inside, the stove is yet unlit. The cold invades the cabins, where water now freezes at night. The Investigators settle into the realization that they could remain stranded in this dangerous position for perhaps eleven months, perhaps the rest of their lives.

A fire hole is dug through seventeen feet of ice. Fifty-seven tons of stone are collected and piled on the beach for ballast. When darkness falls, the lower deck is transformed into a factory. Tailors and cobblers are in great demand. With an urgent sense of purpose, nearly every man busies himself altering new clothes and mending old boots, trousers, and coats. Winter preparations continue until midnight, September 10, when the floe breaks from shore and carries the ship away.

12.

BAY OF GOD'S MERCY

On the south shore of Melville Island, a lone block of sandstone overlooks the barren beach. The anomalous boulder, twelve feet tall and twice as wide, was dropped in place by an ice shelf that pushed over the island some ten thousand years ago. William Edward Parry anchored his ships nearby in August 1819, in a place he called Winter Harbour. At just twenty-nine years of age, he was the oldest officer in the expedition—his second in search of the Northwest Passage.

Parry and his men were the first Europeans to winter in the Arctic Archipelago, and they were stunned by the cold. Determined to preserve the health and morale of his crew, he filled the interminable hours with schooling, exercise, and amateur dramatics. His surgeon, Alexander Fisher, spent not a few of those hours chiselling the names "Hecla" and "Griper" into what would become known as Parry's Rock. For Robert McClure, reaching this rock means the close of a circle of knowledge now thirty years old. Reaching the safe harbour it overlooks—the only one known to exist in this part of the Arctic—is the desire of every man aboard the *Investigator*.

In the early hours of September 11, 1851, the crew scrambles up on deck, but there is little anyone can do. Locked in her cradle of ice, the ship is cast helplessly adrift. Over the following day, every available trick is employed to free the vessel, including the detonation of several hundred pounds of gunpowder—all without success. As if mocking their efforts, the ice splits on its own after they abandon the attempt. The shock throws five men from the deck and onto the ice, where they gather to watch the ship's slow, ungainly slide into the sea. The *Investigator* is eventually brought under control and the stranded men hauled onboard. By the time an anchor line is laid to a grounded floe, they have drifted fifteen miles north along the coast.

Eventually, they cast off again and spend the next few days dodging and weaving between thick chunks and large, old floes. A broadside swipe stalls them for a time, and then a lead opens along the coast. When the route closes in, the men deploy various combinations of sail, pickaxe, saw, and gunpowder to free the ship. In a single day, five hundred pounds of powder are used, half of which is placed in an old rum cask and sunk sixteen feet into the ice. This highly effective detonation launches not a few fish up and into the air. The men blast and weave their way round the cape, and then begin travelling east along the north coast of Banks Island.

Just a few days before, winter's hold seemed complete. Now summer appears to have wriggled free and is on the move again. Ducks, owls, raven, geese, and fox are all seen in the space of a single day. A pair of bowhead whales makes an appearance in a lead half a mile from the ship. Likely a mother and calf, they remain in the vicinity for the better part of an hour, basking near the surface, taking great, audible breaths, arching their backs above the surface before a deep descent. They take turns rising out of the water, head-first, for a view of their surroundings and the ship hemmed in near shore. Sometimes, they rise almost entirely out of the water and then fall with a tremendous splash as if to flaunt their freedom.

The whales gorge on small, shrimp-like crustaceans about the length of a man's fingernail. On this food, krill, they can grow to sixty-six feet—more than half as long as the *Investigator*—with 40 percent of that being taken up by their "bowed" head, jaw, and baleen adapted to exploit their tiny prey. Krill, in turn, feed on even smaller crustaceans called copepods, remarkably abundant at this time of year thanks in part to a relationship between undersea geography and the location of the polar ice pack.

Not far off the west coast of Banks Island, the seafloor falls in a rapid descent from about fifteen hundred feet to nearly ten thousand feet. The ridge also runs along the northwest coast of the continent, framing the deep Arctic Basin. In summer, the edge of the polar ice pack tends to locate directly overtop the continental shelf, where winds blowing across adjacent open sea draw nutrient-rich water up from the deeps, pulling it toward the seasonally abundant light. When this occurs, the water comes alive. The nutrients are consumed by copepods, collectively the largest single source of protein in the sea. In addition to krill, copepods also provide food for fish, which in turn feed birds and seals. Seals become meals for polar bears, wolves, foxes, and ravens. If the polar pack is located too far from the continental shelf, or is absent altogether, the upwelling will stall—along with the fortunes of the Animal Kingdom upon which it depends.

Bowhead whales have been hunted commercially in North America's eastern Arctic since 1611. A large whale yields up to one hundred barrels of oil, which is rendered down into lamp oil, cosmetics, and soap. Also of great value is the whale's elastic baleen—up to fifteen hundred pounds—which is fashioned into parasol ribs, buggy whips, and corset stays. Bowhead whales are slow-moving, stay close to shore, and float when harpooned. Whalers know them as "right whales," as they are considered the right whales to take. Their presence is already drawing more and more Europeans and Americans to risk seeking their fortune on the edge of the Polar Sea.

Ice Master Steven Court is sent out to explore the coastline and returns with news of a possible route. The rudder is shipped yet again and the Investigators are soon underway, sailing a lane of open water, rejoicing at their good fortune so late in the season.

Their confidence is short-lived. A cycle of hope, disappointment, and fear rules the next five days. Open lanes appear and then close, or appear out of reach. Blasting helps move the ship along yards at a time, until she is caught up in the rushing stream. Small chunks of ice are pulled along in the tidal surge, wheeling in the numerous eddies. Larger cakes and floes travel with a force arrested only by solid land. Floes from the polar ice pack arrive singly, and then in overwhelming swarms. It is unlike anything they have previously seen. Even the ship's clear-eyed surgeon is at a loss to explain how the ship endures, short of divine intervention.

He has, however, a ready explanation for their seeming inability to act when a lane does open and offers a route of escape. He is convinced it is the hesitance and indecision of McClure that has cost them several prime opportunities. In his view, the cumulative effect of these failures is beginning to have a "fatal influence" on the voyage. Even Miertsching observes that the captain is making himself sick with the relentless worry and strain.

And still the *Investigator* squeezes between soaring limestone cliffs and jagged, composite floes, heaped thirty feet above the waterline. The continuing critical position of the ship leaves the men with only a few minutes on shore to gather specimens. Fossilized shells are found in abundance, sedimentary remains of a delta as old as five hundred million years.

Heading east, the coastline settles into a series of desolate hills. There is no feature resembling a bay or cove to shelter the ship. Exposed to the full fury of westerly and northwesterly polar winds, floes drive into shore with a velocity that abrades and scars the land. The wreckage of ice is piled on shore as high as a hundred feet. The officers agree: they have encountered nothing to compare. Armstrong, who has travelled to five continents and sailed many seas, is given to neither sentiment nor exaggeration. Having witnessed the scene, he is moved to offer this warning to future mariners—should his journal survive.

Our passage along this part of the coast was a truly terrible one—one which should never be again attempted; and with a vivid remembrance of the perils and dangers which hourly assailed us, I feel convinced it will never be made again.

By 1 p.m. September 23, the ship is sailing at five knots. The men are emboldened with their most hopeful mood in weeks. But then a veil of fog descends. The wind rises from the west and a flotilla of ice arrives, bound for the same destination. On the bow, soundings are taken every other minute. In the crow's nest, the ice pilot directs the course of the ship with frequently changing orders sounded down by means of a trumpet. His broadcast comes to an end with the words, "Ice everywhere." It seems their final move has been made. And yet the following floes push in.

"Hold on!" the pilot cries. "The ice is splitting. It's opening a way for us."

The view from below shows the *Investigator* continuing into a blind lead, headed directly for thick ice just five boat lengths away. The men on deck brace for impact. Then, through some unseen force, the ice shifts and makes way, allowing the narrowest possibility of escape. On either side are high blue walls, seemingly as dense and hard as stone. For Miertsching, it is like the parting of the Red Sea. The ice has opened for a purpose. The ship slips through the gate, and soon they are sailing in lighter ice, more open seas.

The edge of the pack is now plainly visible, along with an open lane leading to the east. As night falls, a gale whips up as wild as they have seen. The storm sends wave after wave of heavy floes while the pack remains offshore. Through failing light and blowing snow, the sailing ice is now a ghostly presence—more felt than actually seen.

The Ice Master is sent to take soundings in the whaleboat, and the ship follows close behind. Despite—or perhaps because of—their intense anxiety, some men predict that Melville Island will be reached by morning. But the risk is far too great. The order is given to shorten sail and wait out the night.

The men have been up top all day and are in need of rest and a meal. After a short break, they return to reorganize the rigging and deck. At first light, they will continue east, across that final, untravelled stretch of the Northwest Passage—the last sixty miles. Despite the storm and the lateness of the season, despite the danger, some men again openly discuss the possibility of sailing all the way home to England. Leaving them to their duties and dreams, the captain retires to his cabin below.

McClure is sitting across from Miertsching, sipping tea, when the ship runs aground. The shock sends both men lunging through the hatchway and back up on deck. The wind continues to drive heavy ice toward the *Investigator*, now trapped on an unseen shoal. If the ship is wrecked here, there is no hope they will survive the night. The soaked, exhausted men now begin shifting heavy items to whaleboats in the dark. But their efforts to lighten the ship are wasted—the *Investigator* remains aground. The captain orders his men in from the cold. As they pass below, he pulls Miertsching aside to request his company as soon as he's changed his clothes.

A few minutes later, the brother finds the captain standing alone in his cabin, with an open book in hand.

"See how holy writ mocks me," McClure exclaims. "In this crisis and extremity, when all our lives are trembling in the balance, I opened the Bible to find words of comfort, and this is how it answers me."

He hands the book to Miertsching, pointing to Psalm 34.

> O magnify the Lord with me,
> and let us exalt his name together.
> I sought the Lord, and he heard me,
> and delivered me from all my fears.

"It is," McClure says, "in flat contraction to our present situation."

For Miertsching, this is the moment to tread carefully, to deflate expectations. It is a moment that requires his most subtle arts. At last, the brother tells McClure that he too has sought answers in random passages of scripture, only to find them inappropriate or lacking. It is true, he says, in times like these, such a passage can be more of a disappointment than consolation. In the fullness of time, however, the meaning and answer will be revealed.

"I thank God that my mind and understanding are unclouded," McClure replies, "and I well know what our situation is."

The captain sips his tea. As they consider their fate, the ship is struck with a shattering blow. Again, both men rush through the pitching ship and back on deck as the *Investigator* rolls and then settles—afloat and free.

At seven o'clock in the morning, McClure is awoken following but two hours sleep. In the predawn light, he orders the cargo brought in from the whaleboats and stowed in preparation for the final dash across the strait.

As morning progresses, the fog is driven off by a strong west wind, and the truth of their position is revealed. They have arrived in the cove of a large bay bordered by a sandbar that repels the floes. Beyond the sailing ice is a clear view of Melville Island—tantalizing in its proximity. They have endured so much, and have been so clearly favoured. Can they risk one more day's battle with the ice? Not, McClure decides, without rest. The ship is anchored and the crew lies down to a twelve-hour sleep.

The following day dawns cold and bright. Even the wind has calmed. McClure walks over the newly forming ice, in search of clarity.

When the ship's company is assembled on deck, the captain thanks them for their devoted service in the face of recent trials. They have accomplished much this summer. Banks Land has been proved to be Banks Island. The whole of its coast has been thoroughly examined—without the slightest trace of Franklin. A *second* Northwest Passage has been discovered, although treacherous beyond belief, and they have survived this passage without the loss of a single life. There is much to be grateful for. But they will go no farther this season. He will not risk their lives in a run across the strait. They have arrived at what appears to be the only safe harbour on the island's north coast, and for this they should be grateful. The ship will be made ready for winter. To ensure that their provisions last, however, sacrifices must be made. Rations will be reduced by a third.

In acknowledgement of his gratitude, McClure christens the sandbank and cape "Point Providence" and the sheltering harbour "Bay of God's Mercy."

Armstrong finds little comfort in these touching names, or in their winter position. The surgeon notes a gloom settling over the crew. He sees their predicament—and the decision to abandon their course with victory in sight—as proof of something less than divine favour. His attention is drawn to the bright floes sailing east toward Melville Island, a mere sixty miles away. Why didn't they push along the pack edge through the loose ice and force their way to Parry's Rock? He is convinced that had McClure found the nerve to take bold action, they might well have made it. Perhaps the ship would have been forced to winter among the floes. While far from ideal, they have already proved this is possible. At least they would have been that much closer to the goal. For the sake of his own honour and ego, to inoculate himself from culpability, the surgeon makes his opinion known and then records it in his journal.

Entering this bay was the fatal error of our voyage.

Among his fellow officers, Armstrong's popularity has not improved. In this sober assessment, however, he is not alone. While some men are content to count their blessings, others conclude that perhaps this bay is properly named—but not for the reasons intended.

"It would have been a mercy," they say, "had we never entered it."

two

13.

HUNTERS, SCAVENGERS

October 1, 1851

The daytime temperature in Armstrong's cabin is eight degrees below freezing. Crystals form on the ceiling and walls, transforming the space into a glistening grotto. The flame of a candle and his own body heat are enough to induce a kind of rain, ruining any imagined sense of comfort. At night, hoarfrost collects on his beard and blanket, frequently bonding the two together. Some days, the surgeon is unable to keep his journal because the ink freezes in his pen. The cold is at its most invasive at this time of year, the final days before the hatches are closed, the stoves lit, and the ship dressed in her winter housing. In the meantime, he is forced to remain in constant motion. While exercise offers temporary relief, it cannot compare with the allure of his berth. Sleep is the only reliable reprieve from constant cold and hunger.

Outside, the ice is already six inches thick and forms a smooth field around the ship. Four hundred paces in one direction lies the beach; six hundred in the other is a large block of multi-year ice, which is mined twice daily and melted down into drinking water. Nine miles long by four miles wide is the determined shape of Mercy Bay.

Ventilating tubes are installed, and the ship is sealed at last. Red-hot cannon balls are carried throughout the cabins and decks in an attempt to combat the damp. Despite these measures, condensation drips between decks—soaking clothes, bedding, and skin. An inventory reveals that yet more food has spoiled.

Armstrong grows increasingly worried. So far, the men are healthy, but this state will become difficult to maintain on the following rations:

6 oz. bread
6 oz. preserved & salted meat (beef & pork on alternate days)
4 oz. peas
2 oz. suet
1 oz. tinned vegetables (potatoes, carrots, barley, rice alternately)
1 1/2 oz. sugar
1 1/2 oz. rum
1 oz. lime juice & sugar
1 oz. pickles
1 oz. chocolate (cocoa)
1/4 oz. tea

—approximately 1,500 calories a day. This amounts to well below half of what a sedentary adult male needs while in cold weather; less than a third of that required when he is engaged in moderate-to-heavy physical activity.

The surgeon is well aware that the men need more food—not less—in order to fight the cold and stave off the effects of scurvy. He suspects their bodies will soon begin to waste. While he himself is hungry, the state of the crew is a more pressing concern. Along with the rest of the officers, he has his own private stock of food and benefits from a larger, more varied diet. Sailors have no personal resources.

Ice Master Court and his sledge team return from a five-day expedition along the coast to the east. The goal of reaching the cairn erected last spring by Lieutenant Creswell would close their circle around Banks Island. The officers gather to hear his report. Their journey was not difficult, Court explains, and they spotted numerous caribou and wolves along the way. However, when they reached their destination, they found that a landslide had destroyed the cairn. By far, their most significant discovery was water—open water—spotted a mere eight miles from the ship and extending north and east along the shore as far as the eye can see. Beyond, a water sky.

For Armstrong, the pride of vindication is quickly quelled by disappointment. If only McClure had followed the easterly current to these open lanes, there would be no need for rationing or resentment. Open water! By now they might well have passed Winter Harbour on their way to Baffin Bay and then home to their prize and glory. Instead, McClure followed his instincts and they remain hungry and trapped in a place he names for God's mercy. Further discussion or speculation is pointless. The officers turn their attention from dreams of what could have been to the immediate challenge of survival.

Despite the various setbacks, their current situation is a marked improvement over last winter's locale. The surrounding land appears to offer plentiful game, and most everyone is keen to take up the hunt. Ice Mate William Newton is the first to kill a caribou. One hundred and sixty pounds of fresh, fat-covered meat is butchered and hung on the yards. This bounty is put aside for a feast on Christmas Day.

The Investigators correctly identify these animals as reindeer. They are in fact the same species as the wild and semi-domesticated reindeer found in Europe and Asia, where they have been herded by Arctic and Subarctic peoples for centuries. In the New World, reindeer are widely known as caribou, a name likely derived from the Mi'kmaq word *xalibu*, which means "the one who paws"—referring to the way they dig through snow to expose their forage below. Throughout the year they graze on grasses, willow, sedges, and lichens. They are partial to saxifrage, and their muzzles are often stained purple when flowers are in bloom.

Like all Arctic mammals, caribou are adapted to conserve heat. The hollow guard hairs of their thick white coats trap warm air close to the skin and form a superior insulating barrier against even the harshest winds. Their large, concave hooves splay out like a set of sharp-edged snowshoes, providing firm footing on ice and rock, and serving as effective swimming paddles for crossing rivers and channels in summer. Adults of both sexes grow antlers. Males use their enormous racks to fight for and establish dominance during the rut; females use their smaller antlers for defence during pregnancy and calving. Antlers are shed annually, leaving the bleached white chips and flakes of countless generations littered across the landscape.

On mainland North America, explorers have observed caribou ranging across the top half of the continent in massive herds of over ten thousand individuals that darken and transform the landscape. They migrate as far as three thousand miles each year, the farthest of any land mammal. The subspecies of caribou found in the western Arctic Archipelago is much smaller and paler than their more numerous and wide-ranging relatives to the south. Here on Banks Island, males average only about 240 pounds—less than half the weight of large mainland bulls. These smaller caribou travel in groups of between four and twelve individuals and generally live their entire lives in an area of less than a hundred square miles.

Human beings have been hunting reindeer for many thousands of years, as did Cro-Magnons and Neanderthals before us. Since arriving in North America, the Inuit and their ancestors have relied upon *tuktut* (caribou) for food, shelter, tools, and clothing. Despite this sustained hunting history, wolves remain the caribou's principal predator, thinning and shaping the stock, ensuring that only the fit and fortunate survive.

The Ice Mate's successful kill emboldens both officers and crew. Several week-long hunting excursions are planned to help add to their dwindling provisions. Court, Creswell, Piers, and Wynniatt each take sledge parties out on the land. Numerous hares, ptarmigan, foxes, even a few muskoxen are seen. Anything that moves becomes a target, but the rate of success is low. Above all, the object of their search is caribou.

One is shot by a hunting party but must be left out on the land overnight. The successful marksman leaves his musket laid overtop the body as a warning to any would-be competitors that this meat is spoken for. When the men return the following morning, they find five wolves feasting on the carcass. The wolves flee as the men approach to curse and collect what remains. The thieves picked up the musket and carried it away so it would not interrupt their feasting.

At night, the hunters lie in their tents shivering, listening to the wolves' plaintive howls. Inside, their breath glistens in the light of the lamp and soon every surface is covered with frost. They sleep three to a blanket bag under stiff raccoon skins that hold little in the way of warmth. Frozen boots must be placed beneath the sleeper's shoulders in order to soften the laces. Come morning, they are pulled on stiff over chaffed feet that continue to sting until numbness sets in.

The hunting parties return to the *Investigator* with thirty hares, twelve ptarmigan, and four caribou. All game is added to the larder, but the ration does not increase. Fresh meat, however, is now issued once a week in lieu of salted and canned provisions. Recently fattened on summer and fall tundra, caribou meat is prized above all. The rich and flavourful flesh provides a noticeable rush of energy.

The anniversary of their discovery of the Northwest Passage is celebrated with extra alcohol, followed by a night of animated conversation and uproarious laughter. The men will eagerly drink to that—or anything else, given the chance.

November arrives with a pack of twelve wolves. They press in on the *Investigator*, howling every night. The bravest member comes right up to the vessel to play with the dog, which follows his newfound friend away from safety seemingly without a thought. Each time, the wolf is keen to lead him back to the others, waiting just beyond musket range. But Mongo always hesitates, stopping short of joining the pack. This game plays out night after night. The men are convinced that, should the dog find himself outnumbered, the wolves will tear him apart.

Mongo is not the only pet onboard. A fox has been kept since May. Old Adam, as he is known, has become so tame that he allows anyone to touch him. He happily takes food from sailors' hands. He seems perfectly happy with his quarters on deck until one morning he is found missing, perhaps scared off by wolves. Mongo, who was always jealous of him, seems pleased to have the men's undivided affections again.

On November 6, the sun shines for ninety seconds and then vanishes for the rest of the year. The melancholy crew is temporarily cheered by the captain's gift of a blanket for every man onboard. But then the rationing extends to light. Last winter, each man was allowed three tallow candles per week. This winter, three must last a fortnight. Candles were considered expensive in Honolulu—now they are invaluable. Some officers have a stock to last the season; others are not so fortunate. One trades a bottle of rum for three candlesticks.

Many men have no goods suitable for barter. A system of commerce is established in which gun wads are circulated as a kind of IOU. One side is marked with the initials of the officer who contracts the debt; the amount owing is shown on the other. Passed from hand to hand like Bank of England notes, they become the currency of all transactions onboard the *Investigator*. Like any monetary system, this one is based on faith. In this case, the faith that these slips of paper—and the debtors named—will eventually find their way to British soil.

Meteors are seen both day and night. On November 24, the star Capella is visible at noon. Aurora borealis frequently smoulder and blaze. The men seek light and hope wherever they can be found.

The distraction of hunting is no longer possible. It is now too dark to see or track game. For exercise, the men are left to pace back and forth on the ice in small groups, earnestly discussing the cherished past or enduring hopes for the future. They try to lift each other's spirits where possible. Old stories are trotted out, retold, and then wildly embellished to make them entertaining time and time again. The subjects remain the same: home, when and how they will return, the fate of the *Enterprise*, and the effects of the reduced ration. Franklin is now rarely mentioned. Most men believe that he and his comrades have long since perished and that any hope of finding them alive is past. They are left to concentrate on their own predicament and fate. Eventually, they shuffle aboard, stiff and shivering. McClure decides this would be a good time to distribute new winter boots from the ship's stores. Another gift, he says, from their most benevolent queen.

A winter storm hits on December 1 that erases the memory of any previously seen. It lasts five days and blows with hurricane force. Just before it arrives, Armstrong records a wild spike in the barometric pressure. During the height of the storm, the men peek out from underneath the ship's housing and see a lone wolf staring back just a few yards away. Like all wolves before him, he runs off before the men are able to aim a musket.

When the fury finally exhausts itself, the men emerge. A snowdrift has accumulated around the ship—thirteen feet high and 165 feet long. Nearly every part of the ship is cloaked in white. The weight of the snow has caused the ice near the ship to slightly buckle and sink. Outside, no serious damage is found. Inside, however, the captain announces that another large amount of provisions has been found spoiled. From this day forward, greater economy and austerity must be practised. It pains him to announce that they must gradually reduce their ration from two-thirds to one-half. This emergency provision will only remain in place as long as they are idle, during these months they are frozen in and no heavy or strenuous work is required. Of course, when they are again under sail, the two-thirds ration will be restored.

The wolves resume their attempts to try to lure Mongo away in the black of night. The howling pack seems to mock the men's carefully apportioned plans, their hopes and schemes. When the wolves are absent, the foxes move in and boldly roam the deck. Even Mongo no longer troubles himself to chase them away. Perhaps they sense new opportunities, or in their own hunger have lost all fear of men.

On Christmas Eve, Second Lieutenant Creswell and Ice Master Court take it upon themselves to decorate for the holiday. They make lampshades with pictures of cattle and sheep from an old edition of the *Illustrated London News*. To the amusement of all, they spread their theme throughout the mess with handsome portraits of steers and cows from a livestock show. Perhaps, with a little inspiration, whatever wild meat they're served tonight will transubstantiate into well-bred beef and mutton.

For men subsisting on half rations, the actual meal of caribou needs no embellishment. This is a day to forget their want. Each man has his generous portion cooked to order, roasted or stewed, with preserved vegetables and puddings on the side. The boatswain's mate declares it is the best meal he has had in his twenty-one years at sea. This sentiment spreads with the endless flow of wine and spirits—to the disgust of Miertsching and the delight of McClure. This one, holy night, they gorge and drink their cares away. Caught up in the goodwill of the season, the captain trades drinks with his men and joins in their rough contests and games.

At muster on New Year's Day, 1852, the temperature is minus-50 degrees. In front of the entire company, McClure engages in a bitter dispute with Joseph Paine. As clerk-in-charge, it is Paine who is responsible for doling out measly portions to hungry men. Since McClure's announcement of the second reduction, his lot in life has been especially grim.

The officers invite the captain to dine with them at supper. It is a moment to rally forces, set the tone, move ahead with a united spirit and voice. Before this meal is over, however, McClure and Armstrong openly clash.

The surgeon is not averse to controversy, or ruining a perfectly good meal. The prospect of continued inadequate rations has been preying on his mind. He would be remiss if he did not voice his concerns. Miertsching considers himself McClure's confidant, spiritual advisor, and friend. He finds Armstrong's insolence incredible. Everyone feels the sacrifice. Is the surgeon unaware of the pressure the captain is under to ensure the survival of sixty-six men?

Armstrong sees his role as uncompromising advocate for the health of the crew. He is unwilling to let another perceived lapse of judgment on the part of McClure go uncontested. Trapped around the mess table, their fellow officers pray for the meal to end.

The men are examined at the end of the month and appear to be in good health, despite a loss of weight and strength. Everyone complains of hunger, and the portions seem to grow smaller by the day.

And then three sailors are caught stealing food set out for the dog. It is a shame in so many ways. McClure is well aware that in desperate times the crew looks to him for leadership, above all. He is also aware that without constant vigilance, more than just discipline will fail. He weighs his response, and then orders the offenders hauled up in front of the entire ship's company. There, the three hungry men are severely whipped for their crime.

If the harshness of the punishment distresses McClure, or if he fears others will judge him ruthless, he dares not betray it. While some will eventually acknowledge the justice, they are unlikely to ever see the great kindness behind this deed. Perhaps this example will prove enough. Perhaps, in the weeks and months to come, it will forestall or prevent the need for executions.

14.

PARRY'S ROCK

A crack of cannon fire tightens the night. The first blast is heard at 8 p.m.; concussions follow every five minutes thereafter. Now and again, rockets blaze up and over Mercy Bay—all without reply. By ten o'clock, three rescue parties are dispatched with lanterns and food for a pair of hunters who have failed to return. Soon, a rescuer arrives with news that the men have been found, but one is in dire straits. Armstrong turns to Piers, orders him to prepare to receive casualties, and then sets off into the dark.

The air is serrated, the wind beginning to rise. About three-quarters of a mile from the ship, Armstrong encounters a group of men. They are in possession of the distraught John Woon, sergeant of the ship's eight marines, and the stiff body of Able Seaman Charles Anderson, the affable and well-loved Negro. Anderson's legs are rigid and his outstretched arms can only be bent with force. No pulse is detected at either the wrist or chest. His glazed eyes remain open, jaw locked. The surgeon grabs hold of the subject and pries his open mouth wide enough to admit a trickle of rum. Anderson, who appears to be in the final moments of life, is placed on the sledge and rushed toward the *Investigator*.

At 1 a.m., the patient is brought aboard, showing no signs of life. When asked what happened, Woon's response is incoherent. The men carry Anderson down to sickbay, where he is laid out on a cot and the surgeons set to work.

Surrounded by inquisitive officers and crew, Woon soon recovers his senses and recounts their ordeal. He explains that Anderson had shot and wounded a caribou, which he followed into the fog. He eventually lost his way back to the ship, panicked, and was overcome. By the time Woon had found him, he had already surrendered to the cold. Despite all means of encouragement and shame, Anderson could not be convinced to get up and save his own life. And then, all at once, he rose. He walked a few paces and fell back into the snow—hemorrhaging from the mouth and nose. This left Woon standing alone in the dark, listening to the distant howl of wolves.

Woon is one of the smallest members of the crew; Anderson, one of the largest. Unwilling to leave his fellow sailor behind, Woon slung both their rifles over his shoulder, pulled Anderson's arms around his neck, and began hauling the him toward the ship. They were forced to make frequent stops. Woon, who grew increasingly frustrated and tired, took to rolling his companion down hills. This not only quickened their pace but also revived Anderson—enough to beg to be left behind. By the time the pair was within a mile of the ship, Woon felt his energy slipping away. He shook Anderson to the point of wakefulness, laid him in the snow, and then set off for help.

In the relative warmth of sickbay, Anderson's feet and hands retain the cold. Armstrong and Piers vigorously rub his muscles and skin. Circulation stirs, then warmth begins to return. A faint pulse is found. Piers looks into the patient's eyes and asks if he knows where he is. Anderson says yes, but is otherwise unintelligible.

The fever and delirium last several days, after which payment for this misadventure comes due. Armstrong is forced to amputate both of Anderson's big toes, portions of other toes, several fingers, and part of his nose. The surgeon notes that the powerfully-built seaman and the slight sergeant were out for the same length of time. Having been forced to haul a much larger man, Woon's rate of exertion was far greater, yet he suffered little from the ordeal. More importantly, Armstrong concludes, Woon did not give up. In this, the ship's ranking scientist finds "striking proof" of the relative physical resilience and moral character of the fair and dark races.

The boatswain's pipe sounds at 6 a.m. Hammocks are hung and the crew cleans the lower deck. At 7:30, they are given dry bread and cocoa. At eight, the officers are served the same inadequate breakfast as the men. And so another day begins. After the excitement of the recent rescue dies away, the routine designed to save them from themselves begins to feel as if it might become their undoing.

At 8:30, Court checks the ship's chronometer as Armstrong and Piers attend sickbay. All muster on deck at nine, after which the men are sent to exercise. They pace up and down along the side of the ship, port or starboard, according to the direction of wind. Twenty or thirty men at a time bundle up with only eyes and noses exposed. Words are seldom spoken. There is no talk of rounders. Instead, each man follows the next in an endless, listless march, staring at the heels of the man before him. When weather forces them inside the housing, their turns are frequent, strides foreshortened. On closed-in days such as these, the wind howls through the rigging above and a fine powder accumulates on everything and everyone. Light provided by a single candle lantern reveals frozen beards, lashes, and scarves. In the best weather—and increasing light—the men sometimes vary their path and strike out for a short walk to the sheltered valleys on shore. Before the pennant is raised at 11:15, the internal clock of each man accurately predicts it is time to return. At 11:25, all hands descend to the lower deck to take their dose of lime juice, and then proceed to dinner at noon. Grog at 12:30, more exercise at half past one—the same monotonous march as before.

At four o'clock, the pennant flies again. Decks are cleared at half past the hour, supper is served at five. At six, the men are ordered to dance and play, but no one answers the call. Instead, they remain in their mess, mending clothes, reading, or stretched out and staring at the beams above. The strains of the fiddle, the thump of boots—even jokes are no longer heard. Hammocks are hung at eight o'clock and the men settle in. In the gunroom, officers play cards or lean over chess and backgammon boards until ten, when they shuffle back to their cabins and pull their doors closed behind them.

By mid-February, the hunt has resumed in earnest. Everyone aspires to caribou. When one is shot, it is opened to reveal a perfectly formed embryo that fits into the palm of a hand. This is recorded as definitive proof that not only do caribou winter on the shore of the Polar Sea, they breed here too.

Among the hunters, competition begets innovation. Sighting a distant herd one day, a sailor removes the red blanket wrapped around his shoulders and places it on his ramrod stuck in the snow. He then lies in wait twenty yards away. Moments later, three curious members of the herd approach to inspect the decoy. Only one survives.

On the Ides of March, the temperature falls to fifty-two below. Despite the cold and hunger pangs, the men roam the land. The faces of the most determined are marked with the boils and scabs of frostbite. Another wayward group of hunters is brought back aboard numb and unable to speak, but this now seems routine.

Caribou are often shot six or eight miles from the ship. When word of a successful kill reaches the crew, a sledge party is immediately dispatched to haul it back, regardless of the conditions. The men are in a race against marauding wolves. Of the nine small caribou killed during the month of March, three disappear in the time it takes for a sledge to reach the carcass.

Recovered from his earlier powder burns, Boatswain George Kennedy shoots a caribou—that manages to get away. Miertsching volunteers to help track it down. Kennedy is the first to reach the animal, which is being devoured by a pack of five wolves. Instead of firing his rifle, he charges the thieves, yelling like a madman. Three of the wolves move a few yards away while two continue to feed. Enraged, the boatswain marches in, grabs one of the caribou's hind legs, and begins hauling it away. But one of the wolves won't let go. It pulls back as the others bare their teeth and snarl. At last, Miertsching arrives on the scene; the wolves relent and retreat. The boatswain gathers up what remains of the caribou, slings it over his shoulder, and marches toward the ship under the brother's protection.

For his efforts—and heroic account—Kennedy is given a small portion of the meat as a prize. The remaining fourteen pounds are surrendered to the common good. Despite the loss to the wolves, this day's hunt turns out to be the most successful thus far. A total of three hundred pounds of meat is added to the larder.

And then a loaf of bread goes missing. It is the second case of theft recorded since the reduction of rations. This time, there is no suspect—no one available to punish.

While out on the hunt one day, Miertsching comes across what at first appears to be a rolling ball of snow. He bends down and snatches up a startled lemming burrowing just below the surface. Resembling a tailless vole, it weighs only an ounce or two and is the smallest mammal in this part of the world. As with most Arctic mammals, its coat changes colour with the seasons, ranging from tawny or mottled shades in summer to pure winter white. With short ears and limbs, and a thick fur coat, its compact body is built to withstand the cold. Still, it is a wonder such a minute creature does not freeze solid in such extremes. It seems impossibly vulnerable staring back from the brother's mitt.

Lemmings range throughout Arctic North America, Greenland, and Eurasia. Two species are found on Banks Island. They feed on the roots and shoots of willow, Arctic cotton, and mosses. Unlike many rodent species, lemmings do not hibernate. As winter approaches, they weave nests of fine grass and rely on snow for insulation. For most of the year, they forage in that gap between the permafrost and snow known as subnivean space, rarely exposing themselves on the surface.

Like the tiny crustaceans in the adjacent Polar Sea, lemmings are a fundamental source of protein for larger creatures on land, providing a critical source of food for Arctic foxes, wolves, snowy owls, and falcons. Both sexes are able to reproduce within a few weeks of birth, and females are capable of raising up to three litters during the brief Arctic summer. Both the survival of Arctic fox pups and the nesting success of snowy owls are closely related to the abundance of lemmings. Their numbers are prone to drastic fluctuations, peaking to plague proportions every four years or so before crashing to near-extinction. In times of overpopulation, they migrate over land, ice, and water in search of new territory. Individuals can even be found on sea ice over thirty miles from shore. In Norway, the spectacle of vast numbers of lemmings stampeding down valleys into lakes and the sea gave rise to a persistent myth that they engage in mass suicide in order to regulate their own population.

Although eagerly consumed by many Arctic animals, lemmings are not a part of the Inuit diet. One of their names for the creature is *kilangmiutak*, or “one who comes from the sky.” A legend common to all Inuit—and Saami of Scandinavia—claims that lemmings fall to the Earth like rain. Among shamans, they are considered a source of supernatural power.

For the men of the *Investigator*, captured lemmings are considered private property. Unlike muskoxen, caribou, hares, and ptarmigan—which must be surrendered to the common good—lemmings are devoured on the spot, still warm, with great relish and satisfaction.

Word spreads that an expedition is in the works. This trip becomes the focus of all enthusiasm and hope. At last, McClure officially announces that he, along with Ice Master Court and six sailors, will make the journey to Winter Harbour. A tent, food, blanket bags, and provisions for twenty-eight days are made ready and loaded on a sledge. Most men suspect one or two of Captain Austin's ships are wintering there, a possibility they have been aware of since the Sandwich Islands. At the very least, a depot will have been left for them. As far as Armstrong is concerned, this is a journey more than a year overdue.

The party leaves on April 11 with high spirits and hearty farewells. Miertsching watches as the men march across their great Northwest Passage, a passage even a simple missionary can see is worthless. Thirty years ago, Parry judged this strait an impenetrable sea of ice. Having the good sense not to wait around for the ice to break, he turned and sailed for home. Observing McClure and his men disappear among the hummocks, Miertsching is convinced that this passage will remain a lost cause as long as the ice endures.

By the middle of the month, the *Investigator* lies exposed, her winter battlements removed or stowed. The old, familiar dream of escape stirs anew. This sentiment is encouraged by the fluted warble of the year's first snow bunting, heard clearly above the silent ice and snow.

Within a week of the bird's arrival, Armstrong diagnoses the first case of scurvy. Others soon follow. The condition reveals itself in loss of appetite, fever and fatigue, nausea and diarrhea, and painful joints and muscles. Armstrong and Piers attempt to ease symptoms as best they can. If good news were to arrive from Melville Island, now would be an auspicious time.

On the afternoon of May 9, a sledge party is spotted to the north. Those not already marching out to offer aid soon assemble on deck to greet their captain's return. When McClure and his men arrive, Armstrong performs a quick inspection. Aside from a few cases of frostbite and snow blindness—and a considerable loss of weight—the men appear free of the dread disease.

McClure informs the ship's company that bad weather was met on their outward journey. When they reached Winter Harbour, there were no vessels to be seen. They quickly located the famous sandstone rock bearing the names of Parry's ships. Atop this landmark, a few stones had been piled over a small copper cylinder. The cylinder contained a note announcing the visit of a sledge party under the command of a Lieutenant Francis Leopold McClintock, a member of Captain Austin's expedition, wintering at Griffith's Island, 450 miles to the east. Dated June 6, 1851, the note explained that seven English and two American ships had sailed through Lancaster Sound in 1850. It went on to reveal that emergency depots were available at Cape Spencer and Port Leopold, six hundred and eight hundred miles away respectively. McClure then composed a note of his own, identifying the *Investigator*'s circumstance and position, sealed it in the cylinder, and replaced it atop Parry's Rock.

In all likelihood, Austin and his ships returned to England a few months after McClintock left his note. As it contained no mention of Franklin or the location of the *Enterprise*, they are left to conclude that the *Investigator* is the only ship remaining on this shore of the Polar Sea. We are, McClure at last explains, left to our own resources.

The officers and crew go about their routine in a fog of shock and grief. Their weary captain seeks out the company of Miertsching.

To the brother, McClure admits that when he realized no help would be found at Winter Harbour, he wept like a child. If only he had dispatched a sledge crew last spring. All hope of help from the east is past, their plight laid bare. Regardless, he must continue to greet his men each day with a cheerful expression, full of determination and hope. Absolution is neither sought nor received. He has the brother's pity.

It is like a curse, this polar ice. Like old age or disease. What if the open water sighted late last year was an aberration? What if the ice does not break this coming season? For McClure, these are questions best borne alone, with courage and dignity.

15.

A TASTE FOR MOUNTAIN SORREL

They come singly, in pairs, and in waves. Snow and Brant geese, common and king eiders, black and red-throated loons, pintail and long-tail ducks arrive to jockey for position as the land emerges from snow. Sandhill cranes, sporting red skullcaps and serious eyes, come to skewer and feast upon the abundance of lemmings. Armstrong is intent on getting a specimen of this large and noble bird. Finally, one is shot and added to the collection: wingspan 4 1/2 feet, weight 8 pounds. Birds come from as far away as Mexico and South America, all with seeming enthusiasm for the vicinity of Mercy Bay.

Even as the air swirls with life, the Investigators feel the weight of imprisonment. The list of men in sickbay grows to thirteen. Heavy snowfalls are followed by fog at the end of May, requiring that signposts and cairns be set up to direct lost hunters back to the ship. When the weather improves, the men hunt by night to avoid the sun. During the month of May, a total of ninety-nine ptarmigan and ten caribou are shot. At this time of the year, most of the caribou are with young, and so yield little in the way of flesh.

In June, snow blindness is more widespread and severe than ever before. The order to wear crepe veils or dark glasses is widely ignored until fourteen victims are added to the sick list, each incapacitated for a week or more. Nearly a third of the crew are now unfit for duty. The rest appear weak and haggard. In response to the sudden deterioration of health, somewhat larger portions of fresh meat are given the men, but this lasts only two weeks. The caribou all but disappear, and hares have become so shy it is impossible to get within firing range. And still the birds continue to arrive: golden plover, purple sandpiper, sanderling, and especially snow geese, with their sweet, flavourful flesh. They are known to mate for life. When one is shot and killed it is quickly propped up on the snow in what appears to be a sitting posture. The hunter waits nearby. It isn't long until its mate arrives, and together the pair find their way into the cook's pot.

By month's end, the temperature climbs to 35 degrees, but the thaw does not progress as the year before. The ice is measured at seven feet, two inches—an *increase* of three inches over the month. At this time last year, there was an observed decrease in the ice of more than two feet. This news further dispirits the crew, for whom hunting in the thickening fog is now forbidden. Armstrong has come to suspect that fresh game is the only thing holding back a more rapid advance of scurvy.

First described by Hippocrates, scurvy is a debilitating and ultimately fatal disease resulting from a deficiency in vitamin C. It has long been the bane of seamen. The British Royal Navy had suspected that scurvy could be treated with citrus as early as 1614, when John Woodall, surgeon of the East India Company, published his handbook *The Surgion's Mate*. In it, he prescribed fresh food in general, and oranges, lemons, limes, and tamarinds in particular. Oil of vitriol (sulphuric acid) would do in a pinch. By 1740, lime or lemon juice was being added to water aboard British ships to make it more palatable, resulting in the improved health of sailors. In 1747, citrus was formally proved as a cure for scurvy by James Lind, a Royal Navy surgeon. By the time Alexander Armstrong took his seat in medical school, Lind's book, *A Treatise of the Scurvy*, was an established text.

Aboard the *Investigator*, the disease continues its advance despite the midday dose of lime juice. Almost daily, a new case is discovered. The men complain of general weakness and depression, and then present with pallor, spots on their thighs, spongy gums, and bleeding from their mucous membranes. Nearly a third of the ship's crew is tainted with scurvy in its various stages. The surgeon suspects this is a result of their general weakened condition and malnutrition—in particular, the lack of sufficient fresh food. A diet of wild game appears to be the only explanation for the avoidance of scurvy by the Esquimaux, for whom lemons and tamarinds do not exist. But now the ship's stock of caribou meat is exhausted, and small game does not keep pace with demand. When the fog lifts, what little meat is brought aboard is the product of many hours spent by weakened men splashing and wading through pools of melting snow. On July 1, Armstrong delivers a full report to the captain. He makes it clear that the men must be placed on a more adequate diet as soon as possible. If not, the outcome could be dire.

"If the present ration is maintained," McClure replies, "our provisions will allow us to survive until next August, if need be."

Next August. Meaning more than a year of continued malnutrition. Armstrong tries again.

"I have presented, as clearly as I know how, the current state of the crew and the growing threat to their health. If the full ration is adopted for at least three months, it will enable the men to better withstand the labour required of them upon the breakup of the ice. Should we succeed in making progress to the east—leading us closer to the depots at Port Leopold and Cape Spencer—then we will have nothing to regret. If we are forced to remain where we are, they will at least have a better chance of withstanding the rigours of another winter which, I fear, will prove fatal to some and severely felt by all."

McClure thanks the surgeon for his considered opinion. Of course there is nothing he would like more than to increase the ration. But this, he claims, is not within the realm of possibility. Hope remains for the breakup of the ice, and their escape, but this is far from certain. As captain, as the man ultimately responsible for the welfare of everyone aboard the *Investigator*, he would fail in his duty if he did not plan for the worst—despite discomfort or disagreement.

The only solace to be found is out on the land, away from the sickness and gloom. McClure often travels with Miertsching, who proves reliably sympathetic company. Together, they seek the higher ground where the snow has melted, the land is exposed, and impatient little flowers bloom white and yellow even before their leaves unfurl. The brother stops now and then to collect grasses, moss, and other plants, which he brings back to his cabin to press, dry, and add to his collection, now totalling 3,785 specimens.

They find long-abandoned caches and rings of stone once used to weigh down the borders of skin tents. On a small island, they discover the ruins of stone pit houses, one of which measures eight by five feet. Its whalebone roof, long since collapsed, lies in shards on the floor. They record these as yet more proof that this shore of the Polar Sea was once inhabited by a whaling people, albeit many generations ago. Mostly, they find piles of stones. Once used as reference points, marking hunting grounds or caches, these *inuksuk* still point the way. They become well known to the Investigators, who use them to navigate back from the hunt—from which they now return empty-handed.

Because game is now so scarce, only the surest marksmen bother. Sergeant Woon still wanders the hills, ready to shoot anything that crosses his path—but especially foxes, as they, like lemmings, are not considered common stock. They are consumed by the hunter. Woon is returning from another unsuccessful hunt one day when he encounters a pair of muskoxen—one of which is lying down, asleep. He has not laid eyes on game this size in many months. He immediately advances to within 120 yards, sights the larger of the two bulls, and pulls the trigger. Now both animals are on their feet. The larger bull seems not to be wounded at all, but instead approaches within forty yards, horns lowered in warning. Woon reaches into his pocket, reloads his rifle with one of the two remaining balls, and fires again—striking and enraging his target.

Now the second bull approaches. Woon loads and fires his final ball, which passes through the first animal's forehead, sending it to the ground. Having no ammunition left, he quickly removes the screw from his ramrod, loads it in the chamber, and fires at the remaining muskox. Now struck in the neck, the bull desperately charges. With no remaining options, the sergeant loads and fires his ramrod, which flies true and enters the bull's left fore-shoulder, passing through the ribcage and lungs, out the right flank, and into the tundra. The beast drops at the sergeant's feet, bringing to a conclusion the most successful hunt in the history of the *Investigator*.

And still these bulls won't go easily. Sledges dispatched from the ship arrive the following day to transport the carcasses back through treacherous conditions. One sled ends up slipping into a patch of open sea. A sailor clings to the sledge and muskox as they float into deeper water. He is immersed for fifteen minutes before being brought ashore with few vital signs. His companions are quick and determined to save him, but it is only with great difficulty that he is revived from the third and final stage of hypothermia. The prize is worth all this and more. Back at the ship, the larger bull is declared the most impressive specimen any of them have seen: 7 ½ feet long, 6 ½ in circumference, 767 pounds—head and hide attached. The younger animal weighs 565 pounds. Once butchered, they add a total of 647 pounds of fresh meat to the ship's depleted stock, and another daring feat to the soaring reputation of the diminutive Sergeant Woon.

For a brief time, Brant geese become easy prey as they have begun to moult and can often be chased down. But travel on land is difficult. On July 17, the temperature is four degrees above freezing and the new snow melts as soon as it lands. With each step, the men sink into the tundra, kept wet by the impenetrable permafrost only inches below. The once-silent land is now filled with a chorus of countless little waterfalls and rivulets coursing over and around rocks and dwarf willow.

As July draws to a close, birds on shore become so wary it is virtually impossible to approach close enough for a viable shot. Hunting on the ice is little better. Ice Master Court and two seamen fall through the floe one day, managing to save themselves but losing three rifles to the sea. Miertsching is ready to give up hunting altogether after returning with nothing more than butterflies for his collection.

And then both he and Piers suddenly succeed. The brother shoots a seal, which he easily retrieves, as the assistant surgeon spots another hauled out and dozing on the floe. Remarkably, Piers is able to creep up on the animal and dispatch it with a single blow to the head with his rifle butt. As with lemmings and fox, seals are not rated as communal game and therefore are not surrendered to the clerk-in-charge. The brother and assistant surgeon are in the mood to share their good fortune, but only with those of the higher ranks. They invite the captain and their fellow officers to dine with them on the dark, oily meat. This windfall goes some way toward relieving the oppression of the cool and foggy weather, and the interminable wait for their release from Mercy Bay.

Each day, a man is dispatched to the sodden shore to hike atop an eight-hundred-foot hill to take long, ponderous looks at the ice-covered sea. Each day, he returns to report that there has been no change, no sign of breakup or movement. This consistent, unwelcome message now greets men who have almost nothing to do. There is scarcely any game left on ice or land. When waterfowl occasionally splash down on a pool near the ship, a dozen men rush out, guns ablaze, succeeding only in scaring the birds away. This year, the wait is unrelieved by ballgames or kite-flying. The men mill about with empty stomachs, bereft of any sense of pleasure or—for some—even the will to live. It is a scene Miertsching takes great pains to escape.

The brother takes long walks on shore, searching the tundra for distraction among tenacious, living things. On one such excursion, he notices a small, dark green plant growing in abundance on the south side of a sandy hill. He bends down, plucks a few kidney-shaped leaves, and pops them into his mouth. Crushed between his molars, they release a burst of flavour, sour-tart, that reminds him of Eurasian sheep sorrel. He has, in fact, discovered a patch of mountain sorrel, a common Arctic plant and a rich source of vitamin C. He reaches into his pocket, opens his handkerchief, and proceeds to fill it with leaves. He carries this treasure back to the ship, carefully washes and then presents it to the captain and his fellow officers as a salad. Both Armstrong and Piers instantly recognize its potential to both cure and prevent scurvy.

Armstrong quickly seizes this new opportunity. Each day, upwards of fifteen men are ordered ashore to gather the herb, bringing back between eight and twelve pounds. The leaves are regularly served fresh at dinner with a little vinegar or mustard. The benefits are clearly observed—the majority of the crew both appear and claim to feel better. For the moment, the advance of scurvy has been checked. Armstrong, however, gives no credit for this discovery to the Esquimaux interpreter, whom he declines to refer to by name and ranks below the ship's most junior seaman.

August 10. The ice immediately surrounding the *Investigator* has been blasted away to ensure the ship is ready to manoeuvre out of Mercy Bay. For the first time, the daily report from the mountain has changed. Out in the strait, some of the ice has been seen in motion, and a strip of open water is visible toward Melville Island. This news is received as answered prayer. But the sense of joy and relief is quickly doused when a member of the crew is caught stealing a loaf of bread. With their departure seemingly imminent, it is critical that McClure not appear unnecessarily harsh. Because his earlier attempts at discipline did not achieve the desired result, he must at least meet the severity of those punishments. He orders three dozen strokes of the lash against the back of the guilty sailor.

One week later, the entrance of Mercy Bay opens for a time, revealing a tantalizing lead extending ten miles to the east along the shore. Forty-eight hours later, the lead freezes over. By August 21, the only remaining open water is a strip along the beach about fifty feet across. To keep the crew occupied, they are made to fish. For three days, they return to the small rent in the ice, casting and hauling their net with the aid of the rubber boat. In total, they catch 173 small fish of an unknown species before even this small hole is sealed.

The atmosphere onboard grows desperate. As the last days of August slip past, the temperature barely struggles above freezing. A few men still point to last year, when winter's hold seemed complete before the ice resumed moving in September. Perhaps, they say, this luck will play out again. When released, maybe they will find a suitable lead to the east, and sail past one or both depots on their way home to England.

The new ice is now sufficiently thick for the men to skate ashore. But there is little point in making the effort, as even the sorrel is covered in three inches of snow. The idle men wander between decks with empty stomachs and heavy heads. Game is neither seen nor heard. Even the music of melting snow has stilled. And then a young sailor, Mark Bradbury, disturbs the night with terrible, unnatural sounds. For some days he has been distracted and confused, and he is now placed under close watch by both officers and crew.

The brother retreats to his cabin for private prayer, but men soon seek him out. They come to hear him read aloud from the Bible, explain the stories in simple terms, and remind those who believe that for God, all things are possible.

McClure avoids company. He leaves the ship to wander the white hills alone, oppressed by anxiety and fear. He assumes responsibility for the daily hike up the hill to observe the ice, and the burden of delivering the disheartening reports when he returns. As much as possible, he attempts to strike a cheerful, positive air with the men. But Miertsching can see he holds darker thoughts. Passing the captain's cabin door at night, the brother hears the sighs and whispered prayers of another troubled soul.

—

At morning muster, September 9, McClure waits patiently on the quarterdeck for the men to find their place and focus. Like condemned men, they stand ready to receive their sentence, steeling themselves for the details, relieved that at least their uncertainty will now come to an end.

"After careful observation," McClure declares, "it is my conviction that the ice will not break up this summer. Therefore, we are compelled to pass a second winter here at Mercy Bay."

In these words, Armstrong finds no revelation, only a fact well known to every man aboard.

"I will do everything in my power to make your lives as pleasant and comfortable as possible. Do not lose heart, but keep firm faith and trust in God, under whose protection we remain . . . I know this is unwelcome news. Discipline yourselves and behave like British seamen, whose steadfast courage has never yet failed. For my part, I have a firm and unshakable conviction that *not one* of us will be left behind. We will all be safely returned to England."

For Miertsching, this speech strikes a delicate balance. He judges it solemn and impressive, the product of careful consideration.

"For a year now, you have not received the full seaman's ration. And still, through God's good favour and the work of our hunters, we find ourselves yet in good health. But now, as we must make our remaining provisions last until next summer, I am forced to make another small reduction in the daily ration. This reduction will remain only for the period of winter inactivity, and thereby ensure we remain in good health . . . I have some personal food that

I will share in common with you on appropriate occasions."

Reduction? *Remain* in good health? The surgeon cannot believe his ears. This impression must be corrected.

For the brother, there is nothing to do but humbly submit.

And then McClure's plan is revealed.

"Come April, thirty of you will proceed homeward, divided into two parties. One party will travel south through the continent, by way of the Mackenzie River. The second party will proceed to Cape Spencer on Beechey Island, where that note discovered at Parry's Rock leads us to expect both provisions and a boat will be found. With this boat, our second party will set out for the Danish settlements on Greenland. Those of us who remain will endeavour to sail the ship next year, if we are so fortunate. If the ice again fails to break, we will retreat up Lancaster Sound to such help as the Admiralty will surely send when they realize our necessity . . . I know that your hunger is keenly felt," McClure concludes. "But our hardships fall far short of those endured by many an honoured Arctic expedition."

An invocation of Sir John Franklin and his lost expedition? Armstrong feels a welling contempt. To propose such an undertaking to debilitated men. Six hundred miles to Cape Spencer and then through Baffin Bay to Greenland in a small boat? The men are unfit for such a journey now, let alone in eight months' time, following the passage of another dark and hungry winter. Having made his mind known on numerous occasions, he now feels compelled to confront the captain, yet again, with their current reality and a sober prognosis for the future.

Following this formal address, Armstrong seeks a private meeting with the captain. He declares the men unfit to perform the service McClure just so publicly described.

"If called upon to make these journeys," the surgeon warns, "make no mistake—this will result in great and inevitable loss of life."

That night, in the solitude of his cabin, this warning returns to haunt McClure until it causes an almost physical pain. Despite what was said to the men on deck today, he knows that unless abundant game returns to Mercy Bay, they will not survive the winter on the provisions that remain. All hands will be forced to abandon ship next spring.

Come what may, that decision will be his alone. No man will force his hand.

16.

THE WASTING

Those that are able have already departed; those left behind make do. Food stocks are down, the young have long since fledged, the days are much diminished. By the middle of September, the great evacuation is past. Some fly south only as far as open water, others well beyond the range of the cryosphere. One species has left Banks Island for the opposite end of the Earth.

Arctic terns are possessed of red feet, white cheeks, and black caps and napes. Their bodies are coloured like the undersides of rain clouds. The feathers of their long, thin wings are translucent near the tips and their tails are deeply forked. Exceptionally elegant and acrobatic seabirds, they enjoy two summers each year and unmatched exposure to daylight. After nesting here and across the circumpolar Arctic, they embark on the longest migration in the animal kingdom. Their annual, 24,000-mile round trip between Arctic and Antarctic seas links the polar regions, extending from the Earth's northernmost land to its southernmost open water. With a lifespan of up to thirty-four years, a single Arctic tern can travel half a million miles, the equivalent distance of a flight to the moon and back again.

Migration allows birds to make the most of seasonally abundant food as well as prime breeding and nesting sites. The tremendous stress and danger of this undertaking are more than offset by the benefits, including larger broods. While the migratory urge is an inherited trait, the ability to navigate is a more complex phenomenon, relying on innate as well as learned behaviour. Like mariners, birds use the sun, geography, and magnetic fields to find their way. But this is only a part of the story. The mysteries of bird migration will continue to baffle and intrigue succeeding generations of observers. While the men of the *Investigator* are loath to see the waterfowl and Arctic buntings go, Arctic terns—with their dagger-like beaks and aggressive behaviour—will not be missed. They were never any use as food.

The ptarmigan of Banks Island have nowhere else to go. They spend their entire lives atop Arctic tundra, ice, and snow. Like other grouse, they have plump bodies with short legs and wings, but their feet are feathered, improving their ability to scurry across drifting powder. Each year, the Investigators observe both rock ptarmigan and willow ptarmigan pass through three colour phases to more closely match their changing and inescapable environment. Their camouflage can be so effective that Arctic fox often walk right past, unaware. Ptarmigan are adept at scratching through snow to reach buried vegetation, and they take advantage of grazing paths cleared by caribou, muskoxen, and even Arctic hares. Confident of their ability to blend in, they often remain perfectly still in the face of danger, allowing keen-eyed sailors within easy musket range. Like all remaining birds and mammals destined to survive the coming winter, they have built up reserves of fat and are in peak physical condition.

Onboard the *Investigator*, the men are increasingly lean. The last game was shot back in July, and this meat lasted until August. Now, aside from the occasional ptarmigan and hare, no fresh food remains.

The daily ration of vegetables is cut to just 2 ½ ounces and now consists mainly of beans. To ensure no nutrients are lost to boiling, the cook devises an ingenious method of preparation. First, hard beans are smashed with a hammer and then carefully poured into the coffee mill. Once the fragments are ground into a powder, a little water is added and then worked into a kind of dough, which is sealed in a bag and baked like pudding. This mass is brought to the table without the slightest hint of butter or fat of any kind. The ration of meat remains unchanged at six ounces, but now the men can't help but notice how, after cooking, it is reduced to a few mouthfuls. To maximize portion size, they begin eating all meat raw and cold. Salt beef and pork are soaked in a little water to draw out the salt, and then instantly devoured as if it were a fine, flavourful roast. The ration of flour for bread remains unchanged at ten ounces, but now tea, cocoa, and sugar are issued in fractions of an ounce. Even the daily dose of lime juice—their last line of defence against scurvy—is cut in half. With the officers' stock of personal food long since exhausted, they exist on the same ration as the crew. The result is an absolute, pervasive hunger both felt inside and mirrored on surrounding faces.

Eventually, the observance of three formal meals comes to an end. In the gunroom, each officer takes his turn doling out rations. After the paymaster delivers the meat, bread, vegetables, and cocoa, the "officer of the day" carefully divides each into eight perfectly equal portions. To ensure fairness, all observe the proceedings. The order in which men choose their portions is determined by lot. The officer of the day is always left with whatever portion remains. Rations are issued in the morning, leaving each man to decide when to eat his share, either as breakfast or dinner—if he can stand the wait. A weak cup of tea is served in the evening. Few men ever manage to save anything to accompany it.

At the end of September, Armstrong and Piers examine the men. They report to the captain that the general state of health is good, given the paltry diet. However, they warn that if provisions are not increased, a threshold will be crossed beyond which are tragic consequences.

Wherever the brother roams onboard, he finds discontent and grumbling. Outside, he observes hungry men burrowing through last year's garbage heap in the hope of finding something to eat. When the snow on the ground was light, it was still possible to ease their hunger with sorrel and grass, but now this forage is lost. And then even the benefit of personal hunting is reduced. At Armstrong's request, an order is issued that, henceforth, small game is no longer the exclusive property of the hunter. Now, half must be surrendered to the sick. Men commonly evade this order by consuming an entire hare or fox raw, out in the field, far from jealous eyes.

This autumn is much colder than the previous two, and yet the vessel remains exposed, the process of housing-in for the winter delayed to allow for the use of the last natural light and the preservation of their dwindling stock of candles. Soon, the combined effects of malnutrition and cold are felt and the sick list swells accordingly. Miertsching's muscles and joints begin to ache. To this misery is added a ferocious, throbbing pain in his jaw caused by a pair of abscessed teeth. These the surgeon are soon forced to pull. The brother shivers all day and then lies in his damp berth, tonguing the gap where his teeth used to be, wondering when this miserable trial will at last come to an end.

His shallow sleep is disturbed by shouts emanating from the adjacent cabin. For several days now, Ship's Mate Robert Wynniatt has been behaving in erratic and puzzling ways. Earlier today, his agitation festered into unmistakable madness with laughing, weeping, whistling, and singing. Like Bradbury, Wynniatt is now under constant watch to prevent him doing violence to himself or others. This situation is all the more disturbing because of Wynniatt's training and access to the ship's store of gunpowder. The surgeons seem to take little notice, focusing their attention on those in sickbay, where their efforts might still have effect. Mental cases are abandoned to their fate among the ship's company.

The noises now penetrating the bulkhead become too much to bear. Despite his aching joints and jaw, the brother pulls back the blankets, slips on his jacket, and approaches Wynniatt's door. Inside, the random shouts and moans seem without focus or relation. The brother knocks and calls out Wynniatt's name. The door opens, but the young lieutenant is not clearly seen in the dark. He waits for the brother to speak.

"My good man, I know you are troubled, but it's very late. For the good of your neighbour, for the good of us all—please, take a little sleep."

After a pause, Wynniatt responds with non sequiturs, revealing the depth of his confusion. Resigned to the futility, the brother turns and shuffles back to his dark berth to await the next volley of invectives or fragment of song.

The weather grows grim, and for three days the men are unable to leave the ship. On October 4, the pressure culminates in a gathering on the quarterdeck. The crew mill about, speaking in low tones until they approach the officer of the watch and demand a few words with the captain. When McClure arrives on the scene, he is approached by four seamen attempting to dispel any appearance of sedition. Their spokesman begins by requesting pardon for the unauthorized assembly. They were left with no other choice.

"The crew has resolved to present a collective request," the sailor declares. "A request for a small increase in the ration. We cannot exist on the present allowance. We cannot sleep for the hunger."

McClure listens as the men clarify and repeat their well-considered pleas. Would he not throw open the doors of the pantry and give each man his fill if he could? He has been hearing similar petitions from the surgeon for some time now. Perhaps they have been in communication. The men confronting him have lost considerable weight, their cheeks are hollowed, their eyes shadowed and desperate. It appears that the hunger is beginning to erode their discipline and cloud their better judgment. The captain must not react out of anger, he must navigate with the greatest of care. He promises to consider their request.

The following week, on the second anniversary of their Great Success, McClure treats the entire crew to a meal prepared with food from his own private stock. To this he adds wine and grog, which has the desired effect. It is the first day in months when good spirits prevail aboard the *Investigator*—save for the confines of sickbay and the minds of the lunatics, who sleep all day and spend their nights in fits of rage and wailing.

Miertsching feels compelled to visit Wynniatt, holed up in his cabin. The brother sits with him for a time, enquiring about his health and spirits, reminding him of their discovery of the Northwest Passage. The ship's mate gives no indication he understands what's being said to him. Instead, he reaches for the brother's hand, kisses it, and weeps without uttering a word.

Through darkening days, Wynniatt is the source of more troubling distraction. He is like a man possessed. His threats and ravings disturb the other men and subject morale to constant strain. Aside from his servant, Miertsching is the only one Wynniatt will allow to approach him. Miertsching takes to visiting him daily to offer comfort and hope, but his efforts have little effect. Wynniatt rejects all advice as well as medicine sent by the surgeon. Recognizing a moment of near lucidity, the brother asks him what, if anything, can be done to help calm his mind and soul.

"I will not cease," the young officer proclaims, "until both the captain and the surgeon are dead."

Weather permitting, the brother seeks release. He trudges across the ice and snow hungry, without purpose—other than to be away and free. When he returns, the burden seems only to have grown. He retires to his cabin hungry, listening to the terror inside Wynniatt's mind come welling up and out in unpredictable torrents and streams.

In bed, the brother's thoughts return to Labrador, where he recalls having seen the face of hunger. It was contained in the eyes. He remembers feeling pity at the time, but this was theoretical, detached. He did not know what true hunger meant—perpetual, inescapable want. It strikes him that it is only now, after these many miles and years, that he at last understands what it was he was seeing.

The sun forsakes them for the third time on November 7. When the skies clear, the stars are sharp and brilliant both day and night. The carpenters begin building sledges for the planned evacuation next spring, but the men assigned to pull them seem less and less able to manage it. When the first caribou of the season is spotted and killed, it is taken as a hopeful sign. But the crew, dispirited and weak, does not take up the hunt with enthusiasm and drive as in winters past. The weight of responsibility for obtaining meat now falls almost entirely on the officers, who assume their duty despite worsening conditions.

The cold now confronting them is unlike anything they have experienced before. Temperatures drop to minus-65 degrees and, when accompanied by even the slightest breeze, proves beyond endurance. Hunters commonly stumble back to the vessel numb, nearly helpless with cold and hunger, unable to speak in sentences until rest and warmth restores them.

Enclosed for long periods, the men find little to occupy their minds and hands. The winter school for sailors has been cancelled for lack of energy and interest. Both students and instructors find it too difficult to maintain concentration. A few still read, but most now turn to needlework. Both officers and crew take up knitting and crocheting to help shorten the hours with a soothing, undemanding task. At any given time, the delicate, measured movements of knuckles and fists can be seen encircling candlelight.

The captain's promise of increased provisions is kept in the form of an extra ration and grog each week. These the men accept with gratitude and the understanding that it is the best that can be hoped for under the circumstances. On near-empty stomachs, the extra alcohol succeeds in inducing a few hours' cheer.

When the temperature warms to minus-42 degrees, the brother ventures out to hunt and meets with great success. In one day he shoots a caribou, a hare, and two ptarmigan. All of this meat he surrenders to the general stock, the smaller game going to the sick. For his efforts, he is rewarded with the caribou's head, hooves, liver, lungs, heart, kidneys, and a pound of flesh. Encouraged and fortified, he heads out on the land again the following day and shoots five ptarmigan, which he delivers to the sick. His fellow officers also meet considerable success. This allows the crew to eat fresh game every fourth day in lieu of canned and salted meat—now over three years old.

The intensity of the cold continues to astound. While out hunting, the stock of Armstrong's rifle cracks and shatters against his shoulder upon recoil. This event repeats itself with another hunter the following day. Then Miertsching and a sailor spot a caribou with an extraordinarily large rack. When the brother fires his rifle, the gunstock holds but the wounded buck lowers his head and charges. The men stand their ground and the sailor brings down the butt of his rifle between the antlers, cracking the buck's forehead and shattering the gun into three separate pieces. A few days later, the brother falls from a cliff into a deep hollow of snow and sprains his foot. He manages to haul himself out and back to the *Investigator*, but is forced to forgo hunting. And then the cold is such that no one dares leave the ship.

For two weeks, the weather continues clear and biting. Onboard, a sense of foreboding grows. When the supply of fresh meat disappears, the brother is compelled to act. His sprain is on the mend, but his fingers have been frostbitten so many times it has become difficult for him to carry his own weapon. A sailor volunteers to help.

When at last they sight a caribou, the sailor hands over the rifle. Miertsching shoulders it, pulls the trigger, and drops the target in the snow. They rush to the animal and the brother puts his lips to the wound. He and the sailor take turns until the flow is staunched. Warm, nutrient-rich, and vivifying, the blood freezes on their cheeks and beards. Next, they open the first chamber of the animal's four-chambered stomach. They reach into the steaming mass and make a meal of partially digested forage. They return to the ship with news of success—and smiling faces covered in gore.

On Christmas Eve, the weather is so foul no one ventures out. However, the mood onboard is bright. The clerk-in-charge releases larger portions in advance of tomorrow's feast. McClure announces that on Christmas Day, each man shall eat his fill. It is understood that this will be their last Christmas together. Come spring, they will either attempt impossible journeys or their provisions will eventually fail.

Those fit to participate seem genuinely determined to celebrate with joy. Many, however, are unable to take part. Sickbay is now nearly full. Clerk-in-charge Joseph Paine and First Mate Hubert Sainsbury are dangerously ill. The lunatics persist in their states of delirium. Bradbury does not rave as much as before, but remains haunted by irrational thoughts and weeps incessantly. While onboard, Wynniatt manages to contain himself, but when he finds his way out on the ice, he gives free rein to his demons. There, on his endless stage, he taunts and rages against the man who once humiliated him over a broken watch. There, on his battlefield, he has launched no fewer than three murderous attacks upon the person of Captain McClure.

Christmas Day begins with divine service, followed by dinner served on tables decorated with care. The surrounding bulkheads are covered with drawings of perilous floes and undaunted ships sailing safely through. The meal is consumed with genuine gratitude and attention to polite manners. The balance of the day is spent with poems, songs, and skits bathed in an extravagance of lamp and candlelight.

The goodwill lingers through New Year's Eve, when Miertsching instinctively counts blessings. Despite their hardships, the Lord has granted him and most of the crew good health through a trying year. In his darkest hours, he has yet been able to find moments of grace. He is thankful, too, that as their situation becomes increasingly desperate, men have begun to seek him out to hear words spoken for the good of their immortal souls. Some have even begun to show by their actions that they see that better way to live and seek to walk this path. As the brother surveys the haggard and greasy faces around him, he feels a kinship despite all the troubles and trials. A ship is a world unto itself—never more so than when lost to the world it left behind. And yet he believes he is not forgotten by family and friends back home, or among the brethren. Exile aside, he trusts he is yet remembered before the throne of God.

Armstrong allows himself a moment of thankfulness and hope as the last hours of the year slip away. Despite miscalculations and mistakes, despite hunger and disease, God has shown them great mercy. He is mindful of the fact that after three years of service, not one of their sixty-six men has perished. Unless a member of Franklin's expedition is found alive, unless they themselves eventually succumb, their tale of fortitude will surely go down unmatched in the annals of Arctic navigation.

Across the table, McClure is similarly awed by the success of their survival. It is more than he could have hoped for. He too is inclined to credit divine providence for seeing them through so many tumultuous events and countless smaller trials and indignities. Perhaps, despite all contrary indications, they will yet survive. They have come too far, endured too much, for this to be their end. Their predicament puts him in mind of Manoah's wife, the Old Testament heroine who saw proof of God's favour in the sheer fact of endurance.

If the Lord were pleased to kill us, McClure concludes, He would not have shown us these many mercies.

17.

THE SURGEON'S REPORT

On New Year's Day, 1853, the crew submits to examination. Scurvy has spread and now begins to appear in more threatening forms. The men complain of dysentery and faint at the slightest exertion. One morning, a sailor awakens to find that several of his teeth have fallen out while he slept. Others find their legs turning black and blue. Sickbay is now beyond capacity, and hammocks are seen on the lower deck both day and night. Frustrated by his inability to provide the nutrition and warmth his patients so desperately need, Armstrong orders them to stay in bed to at least preserve body heat.

Wynniatt has suddenly calmed. This is taken as a hopeful sign—until he quietly mentions his intention to murder the captain and officers before setting the ship aflame. Given his earlier attempts on McClure's life, these threats are taken seriously. His hands and feet are bound. Along with his fellow lunatic, Bradbury, he again takes to filling the night with maniacal howls and raving.

Thus far, Miertsching has avoided most indications of scurvy. But like everyone else, he suffers from slow starvation. The surgeon has him strip down to his underclothes and climb aboard the scale. One of the healthier members of the ship's company, the brother has lost thirty-five pounds over the course of the preceding year. When he considers the journeys proposed for the coming spring, he permits himself no illusions. If these plans are put in motion, there is not the slightest possibility anyone will survive.

The cold continues to break all previous records of their Arctic tenure or any previous expedition. Temperatures fall as low as minus-67 degrees. Air burns the top of the lungs, and exposed skin freezes in under five minutes. When the wind picks up, heat loss from the skin increases, thereby accelerating the effects of the cold. The hull of the ship itself complains and shrinks from the extremes. Metal bolts, nails, and fastenings are heard to crack as they succumb.

Herds of five to eight caribou descend from the hills to the low-lying shore, but in temperatures more than 90 degrees below freezing, few rifles are up to the task. Springs snap and stocks shatter when discharged. Rifles belonging to Armstrong, Paine, Piers, and Sainsbury are now all beyond use. As with their equipment, the effectiveness of the men themselves is much reduced.

Regardless, they manage to shoot a few caribou. Every part of these animals is now consumed, entrails and hide included. Given the fact that fresh meat is now back on the menu, hunters are again allowed to keep smaller game. Thoughts of feasting on an entire hare or ptarmigan send unskilled hunters into the stinging cold. But they achieve nothing other than frightening game. They make no addition to either the general stock or their own miserable ration and return to the ship wasted, their hunger shrieking within them. Staggering back from one such excursion, a hungry sailor falls on the gangway and breaks his arm.

Inside, the men creep about with dull eyes and energy for little more than sleep. Miertsching observes how low these once strong and imposing men have sunk. Pathetic suggestions of their former selves, they are hardly able to stand upright. As it remains to be seen who will be chosen to go, each man attempts to ready himself for an epic march. The only thing they know for certain is that they are unable to face the prospect of yet another Arctic winter.

And then the officer of the watch reports another case of thieving. This time, the ship's baker has been caught. The man's private chest is searched and found to contain meat, flour, and dough. In the face of overwhelming evidence, he vehemently and repeatedly denies any knowledge of how these items came to be in his possession. Accusations follow denial until all at once his pathetic excuses collapse under their own weight and he is forced to confess in full. This is the fourth incident of stealing food. Each time, the betrayal feels both more personal and easier to understand. McClure seems less willing, or able, to make examples of starving men. He orders the baker hauled up on deck and stripped to the waist. This time, the thief receives only two dozen lashes.

At the end of January, the surgeon delivers his monthly report to the captain. More than a third of the crew is incapacitated, most suffering from scurvy and its debilitating symptoms, along with the effects of starvation. Some of those released from sickbay are still feverish and wracked with dysentery. Both Sainsbury and Paine have worsened and now appear close to death. This assessment, along with Armstrong's predicted outcomes, leaves McClure in despair.

And then both officers and crew take note of the brightening sky. This brings improved hunting conditions despite continuing low temperatures. The returning sun rekindles long-dormant hopes. Some men even dare to dream of seeing England once again.

A month later, the strongest men are given larger rations and sent out in the cold. They are ordered to haul sledges to the frozen beach, collect volumes of sand, and lay a broad path from the bow of the ship to the mouth of Mercy Bay. This, McClure hopes, will hasten the effects of solar energy and help open a channel this coming spring. Back inside, the ship's armourer has fallen ill and Miertsching has volunteered to assume his tasks. With care and attention, he fabricates three dozen small mess tins for those men destined to leave in six weeks' time. As the brother beats and shapes the metal, as the crew sprinkles sand atop ice seven feet thick and growing, the surgeon marches aft to meet McClure.

As is his habit, Armstrong does not bring welcomed news. Twenty-one men are in sickbay, and the survival of some remains in question. Despite this business with the sand, despite the slight increase in rations these men gain, not a single member of the ship's company is fit for heavy labour. Make no mistake—their steadily wasting bodies and spirits will fail them. With this, Armstrong concludes his speech and marches back to sickbay.

Directly below, in the carpenter's store, Miertsching attempts to focus on the task at hand. These mess tins are made for meagre meals. They will add yet more weight to heavy loads hauled toward some obscure point far beyond the horizon. In dim candlelight, he pounds out another matching bowl. How many men will survive the coming trial? Not one, the brother suspects. Not a single one.

Two days later, Armstrong is summoned back to the captain's cabin. For nearly six months the ship's company has been aware that they will divide into thirds: one group heading east, one south, and one remaining with the ship. What is still unknown is the composition of these various parties. The time has come, McClure explains, to distinguish the strong from the weak.

"As ship's surgeon, I leave it to you to make the necessary selection."

At morning muster on March 3, the ship's company is instructed to remain on deck to receive the orders for which they all have been waiting.

"An inventory has revealed that our supplies will last only until November. This would be for the entire crew. Because it is neither possible nor practical to abandon ship all at once, I will dispatch only enough of you to ensure that provisions for those who remain will last until the following spring. This will allow for the possibility that the ship again fails to get free of the ice.

"The first party will head to Cape Spencer. There, they will find a boat and a hut containing food, clothing, and coal in abundance. From there, they will continue toward Greenland in the hope of meeting a whaling ship in Baffin Bay and, ultimately, passage back to England. Under the command of First Lieutenant Haswell, the following officers and crew will depart: Assistant Surgeon Piers, mates Sainsbury and Wynniatt . . ."

The names of twenty-six sailors are called out, most of them frail and infirm. After the final name is heard, the reality of the plan sinks in: a six-hundred-mile sledge journey followed by a voyage in a small boat through the legendary bergs of Baffin Bay. Of the officers named, one lies close to death, another is a raving madman.

"A smaller party, under the command of Lieutenant Creswell and assisted by Mr. Miertsching, will return to the Princess Royal Islands and that depot laid by us two years ago." He reads out the names of six accompanying sailors. "These men will recover the provisions and boat and await the breakup of the ice. From there, they will travel south, by boat, along the continental coast and up the Mackenzie River to the Hudson's Bay post at Fort Good Hope. They will continue—with the help of friendly Indians—through the wilds of North America to Montreal and New York. From New York, they will take the next ship back to England."

Montreal. New York. England. Points fixed and knowable, separated only by the space of a few words.

"The first party will take provisions for forty-five days, the second will take thirteen. For a month prior to departure, travellers will receive a full ship's ration to fortify themselves for the journeys to come. They will be excused all other duty and supplied with candles to prepare their clothing. Both parties will depart on April 15."

The standard for selection is now plainly clear: the fit and healthy will remain with the ship, and the sick and weak will be sent away. Sixteen of the men ordered off the ship are currently on their backs in sickbay. Some of those ordered to leave confess to visions of their corpses left lying atop some unnamed stretch of ice to be torn, digested, and ultimately expelled by wolves. Still, some of those chosen to remain with the ship would prefer the chance to flee. They are haunted by the prospect of spending another winter, their final winter, locked in the ice of Mercy Bay.

None of this comes as a surprise to Miertsching. In confidence, McClure had confessed the plan to him long ago. He has had time to come to terms. Over these last few months, the brother has been busy preparing his clothing and personal kit in the privacy and warmth of the captain's cabin.

Following McClure's address, the men crowd the hatch and disperse below. The brother joins the stream and makes his way down to Wynniatt.

"You will be leaving the ship in a few weeks' time," the brother explains. "You and half the crew will be travelling under First Lieutenant Haswell. This is your chance to escape, to return home to England . . ."

The brother repeats himself a few times, rephrasing and reshaping the story. But he sees no sign of comprehension. Similar attempts fail with Bradbury. He too is devoid of understanding and must now be handled with gentleness and patience, much like an idiot child.

Armstrong quickly withdraws and closes his cabin door. He sits at his desk and pulls out a sheet of paper. He warms the inkpot between his palms until liquid begins to flow. Following what he believes are the dictates of conscience and duty—and to absolve himself from any responsibility—he makes an official record detailing his opinion of the captain's desperate plans. Together with his assistant surgeon, Henry Piers, Armstrong is convinced that the men are wholly unfit to undertake such audacious journeys. He records his concerns, then stores the letter in case a higher authority one day calls into question the final decisions made aboard HMS *Investigator.*

As far as McClure is concerned, a captain's reasons are his own. His plan and decision were not made in consultation with senior officers. He does not expect to be popular or loved. Why dispatch the sickest men while retaining the strongest, the most likely to survive? Increasing provisions for the entire crew is not an option, nor is allowing all hands to remain onboard where everyone will surely starve. The sick will not survive the summer. On this point even the surgeon must agree. With this plan, the captain has announced to the more vulnerable half of his crew that he is no longer able to protect or provide for them. This admission does not come easily. If they go, at least they have a chance of saving themselves. It is, McClure believes, the best chance they have—and the only one in his power to give them.

For superstitious men in need of a sign, the killing of a wolf is welcomed news. The seemingly impossible occurs when a party is dispatched from the ship to recover a caribou shot the previous day. When they arrive on the scene, they find a large, pure-white wolf feeding on the carcass. The men fire their guns and miss, just as each time before. The wolf runs off, unscathed and out of range. But this time Sergeant Woon and his men lie down and wait behind a nearby bank of snow. When the wolf returns to feed, Woon carefully sights his target, holds his breath, and squeezes the trigger.

The ball strikes the wolf directly in the chest, passing through its heart, and dropping it atop its final meal. Within the space of forty-eight hours, Ice Master Court repeats the feat they have been attempting for over two years. The wolves weigh eighty and seventy-two pounds. They are carefully skinned. As a special treat, the meat is actually cooked, and judged excellent by all. Finer palates declare that the flavour resembles—but is superior to—Arctic fox, and preferable to polar bear.

The temperature briefly peeks above zero, and this brings hope for an early summer. Those selected to travel are now put on increased rations. However, scurvy continues its advance. Virtually every man onboard is affected with the disease to various degrees. Nine of the men still lying in sickbay are slated to strap into harnesses and begin pulling a sledge in two weeks time.

April opens with snow squalls and strong northerly winds. The new sledges are tested and stowed. The travelling parties spend their time preparing clothes and writing letters to family and friends. They are told that any personal property left behind must now be packed, addressed, and surrendered. Should the *Investigator* find her way back to England, all will be returned. In this dim hope, the brother organizes and packs his thousands of specimens—sprigs of lichen, blades of grass, flattened flowers and butterflies. Regardless of the surgeon's official and more proper collection, Miertsching hopes his own modest effort may yet add to scientific knowledge of the Arctic regions.

And then the captain's latest order is read aloud:

> *All journals, sketches, charts, and other documents, written, drawn up, or prepared in connection with our expedition must, by April 5, be sealed, addressed, and delivered to the captain, who will turn them over to the Admiralty whence they may one day be recovered.*

The brother's journals weigh next to nothing. He would gladly surrender any other item for the sake of these precious notes. His journal contains little information of strategic importance. It is at best a collection of thoughts and impressions. It is the one article that is truly his alone. And yet he humbly submits, trusting in the good faith of his captain and friend—and the eventual release of the *Investigator*.

The increased ration appears to be having an effect. Armstrong notes an improvement in the physical appearance of the crew. Faces are fuller and more animated, the ubiquitous blank stare less pronounced. At times, he can detect moments of levity. Although the sick list has been reduced, some of those who remain are gravely ill and require constant attention. To help with the numerous tasks, he drafts a new sickbay attendant.

Able Seaman John Boyle is himself suffering from dysentery, but seems otherwise capable of service. On April 5, he completes his second shift in sickbay and then lies down in his hammock. Following a short and cheerful conversation with his fellow sailors, he makes a slight movement—as if to rise—and then slumps unconscious. His mates are unable to wake him. By the time Armstrong arrives, the able seaman is dead.

Word of their first fatality strikes like a broadside wave. This first shock is closely followed by the rumour that the tragedy was self-induced. Someone suggests that the man might have taken unprescribed medicine and that this caused his death. Armstrong fully investigates the accusation and determines it patently false. Boyle died of complications resulting from scurvy and starvation.

McClure prefers the sensational version of events. This sailor died through his own mistake, not conditions aboard his ship. Still, it is time to acknowledge the matter, to quickly lift the veil of grief. The men must regain their spirits and prepare themselves to depart in ten days' time.

McClure holes up and prepares an eloquent address. First, he will inform them of Boyle's unfortunate mistake. Next, he will call their attention to all they have accomplished and endured together, the successes attained, the challenges mastered. They will turn sorrow into conviction. Divine providence has seen to their needs thus far and will not abandon them as they set out for home. He will conclude with a reminder that the blackest hour always precedes the dawn, that even the darkest cloud is silver-lined. He will strike the right tone. This death will not cause them to falter, but urge them to survive.

With the ship's sad company mustered above, McClure climbs the ladder to face them.

Those on their backs in sickbay hear only the footfalls of their able-bodied mates, the sighs of fellow invalids, and the shapeless rise and fall of a voice they know to be McClure's.

They do not have the benefit of those rousing and carefully chosen words. They know only that they will soon be called to stagger up the ladder and across the deck to finally face the ice.

three

18.

ADVENT

April 6, 1853

It is no small thing to dig a proper grave in temperatures twenty below. Snow squalls, both yesterday and today, make choosing a proper spot even more problematic. McClure has put off this task all morning, spending the time instead composing letters to the Admiralty for his two sledge parties to carry. He has written them, of course, as if they will be delivered in person, by his own able officers—not pulled from the pockets of their frozen bodies by uncomprehending Esquimaux. The days are filled with forlorn tasks that require everyone to act as if the end were not so near.

At noon, the seamen pace the ice near the ship in the falling snow. They are unable to attempt hunting in such conditions. The captain leaves the ship and walks beyond their circuit in the company of Miertsching. When the wind diminishes and the snowfall wanes, they set out to locate a burial place. They make fresh tracks on the surrounding floe, discussing the brother's upcoming journey across the continent, all the way to New York. They speak of the letter his party will carry and their duty to make the plight and position of the *Investigator* known to the Admiralty as soon as possible. And then the topic turns to the desperation of those ordered to remain and attempt to sail the ship out of Mercy Bay. They try and speak of these efforts as if they could succeed. Perhaps, McClure suggests, with both luck and divine favour, they will one day meet again.

"Sir, if next year in Europe you neither see nor hear of me, then you may be sure that Captain McClure, along with his crew, has perished and lies unburied, but wrapped in the fur coat which you gave me, enjoying a long and tranquil sleep until awakened on the day of the resurrection by the Redeemer, in whom is all my hope and trust."

First Lieutenant Haswell approaches from the ship. He comes to assist in the choice of a gravesite and to discuss preparations for the upcoming funeral. Before he reaches McClure and Miertsching, a sailor is seen in urgent pursuit. Haswell and the sailor arrive together. When the seaman recovers his breath, he reports that an object has been spotted out on the heavy ice, moving in their direction. Everyone's gaze follows the sailor's outstretched arm across the strait, where a dark spot appears to be moving. Clearly, it is a living creature, perhaps a muskox that has lost its way. A second sailor now comes off the ship and toward the men, breaking into an open run. No one has seen a member of the ship's company move this fast in too many months to count.

"They're men!" the sailor shouts. "First a man, and then a sledge."

The captain and the brother look at one another, and then at the face of the newly arrived sailor. All at once, the group sets off across the ice.

The approaching figure walks at a clip. Perhaps it is only a member of the crew trying out a new coat made for the upcoming sledge journeys. However, after all this time together, they know the shape and gait of each member of the ship's company. In this case, neither fits. It has been twenty-one months since they last met another human being. The brother summons the Esquimaux words for greeting. He feels a new hope take root within. He wonders where these people have come from, and if he might go with them.

As the captain, brother, lieutenant, and sailors close the gap, it appears that the stranger is wearing Esquimaux attire. And then the figure throws up his arms in a gesture of greeting. A shout is heard, too distant to discern. They march on with purpose, the brother straining to hear some word or phrase for translation. He is distracted by the surging beat of his heart and throbbing pulse in his head.

"I am Lieutenant Pim of the ship *Resolute,*" the stranger is heard to say.

These English words pass through the men with a flood of adrenaline. The Investigators stand transfixed, unsure whether to trust their senses.

As the stranger approaches to within a few dozen yards, they see an English face blackened with the smoke of a cooking lamp.

The Investigators have no words, each man choked with emotion. They rush the stranger without speaking, seeking only to touch, to grab his hand in confirmation.

Pim is followed by a pair of Englishmen hauling a sledge. When they arrive, the silence is broken. Multiple, baffled, joyous conversations ensue. Questions and answers fly back and forth as the men begin the process of reacquainting themselves with that lost, civilized world. McClure's relief is profound. It is as if the course of destiny has suddenly reversed itself. Possibility and future have opened once again. In this happy confusion, he realizes that these strangers have arrived just in time. They have saved the lives of thirty men who—on his own order—were about to set out on journeys they could not possibly hope to survive.

Back on the deck of the *Investigator*, the carpenters drop their tools. They step over the half-built coffin and squint into the distance. Behind them a stream of men pop up the hold and crowd the upper deck. The distant objects appear to be a party of three strangers engaged in conversation with the captain and his men. Officers and crew line the gunwale, voicing their opinions and fears. Some declare these strangers to be Esquimaux, some say Europeans. Others pronounce it a mistake or a joke, rejecting or unable to comprehend the truth of what they see. When the figures turn and start marching toward the ship as a group, the men erupt in a spontaneous cheer. Those still below, both able and weak, now rush the hatchway to gain the view of what's unfolding in Mercy Bay. Some break out in whoops and shouts, others silently weep. Those still in doubt rush across the ice to meet the strangers. They remain unsatisfied until they themselves can touch them and hear the voices of fellow countrymen, the first they have heard in nearly three years. They surround and receive Lieutenant Bedford Pim and his two fellow officers, Robert Hoyle and Thomas Bidgood. The Investigators have long been convinced they were the only Europeans still alive in the Arctic—Franklin and his unfortunate men included. They are unable to find words for their profound gratitude and relief. Instead, they offer three exuberant cheers.

When the uproar subsides, Pim gives an account of how he reached Mercy Bay. On the arrival of the ships *Resolute* and *Intrepid* in the region last autumn, they found Winter Harbour frozen over. They anchored instead at tiny Dealy Island, forty miles to the northeast along the coast of Melville Island. They began to establish depots in advance of sledge parties to be sent the following spring in search of Sir John Franklin and Captain Robert McClure. Back in England, the father of the *Investigator*'s own Lieutenant Samuel Cresswell—an influential and wealthy man—has been loudly reminding the Admiralty that Franklin and his men are not the only English heroes missing in the Arctic.

On one of these excursions, a sledge party stopped at Winter Harbour. One of their officers located Parry's Rock. He saw the pile of stones on top but decided against disturbing them. Instead, he began to chisel a record of his own visit on the face of the rock. This disturbed the cairn above, and a copper cylinder tumbled into the snow. Believing it to be a relic left by Parry thirty years ago, he thought it best not to risk damaging it in the cold. Luckily, curiosity got the better of him. He opened the cylinder and pulled out a scroll. The enclosed covering cracked and fell apart in his hands. He then broke the seal and found McClure's account of the discovery of the Northwest Passage, along with the position and condition of his ship. The note was dated April 12, 1852—exactly six months before.

On receipt of this revelation, Captain Henry Kellett planned to send a search party to Mercy Bay at the first opportunity. On March 10, Pim's party left the *Resolute* and walked straight into a blizzard. Later on, a broken sledge forced him to send back half of his men. The remainder arrived today, after a difficult and dangerous journey of 170 miles and twenty-eight days, the earliest Arctic journey of its kind undertaken in British naval history.

To the Investigators, Lieutenant Pim's greasy, smoke-blackened face is that of a beloved saviour. To Pim, the gaunt faces staring back present a vision of doom. He and his companions have come prepared for a mission of mercy and relief, and yet the state of these starving, enfeebled wretches is a sight for which they are unprepared. It is an image that reduces them to tears.

And then they are welcomed aboard the *Investigator* to meet those men too weak to rise.

The ship's company gathers and listens to Pim's thorough briefing. In the year 1852, five ships departed England for the Arctic under the supreme command of Sir Edward Belcher. He took direct command of two ships, the *Assistance* and the *Pioneer,* and sailed north from Wellington Channel, where the *North Star* remained as a depot. Captain Henry Kellett was left in command of the other two ships: the *Resolute* and her tender, the *Intrepid.* Kellett's is a name well known to the Investigators. He and his men were the last Europeans they saw. Three of Kellett's men volunteered to join the crew of the *Investigator* at Kotzebue Sound on July 31, 1850. At that time, Kellett was captain of the *Herald* and had urged McClure to await his consort, the *Enterprise.* McClure disregarded this advice and pressed on into the ice. Since then, Kellett returned to England and is now back in the Arctic aboard the *Resolute.* After parting with Belcher at Wellington Channel, Kellett sailed west toward Winter Harbour, settling instead for Dealy Island. And this is how

Pim, Hoyle, and Bidgood have come to find themselves standing aboard the *Investigator*.

These guests, in return, receive an account of the *Investigator*'s Great Discovery, as well as myriad tales of terror, hardship, loneliness, despair, disease, cold, hunger, and madness.

The following morning, McClure selects seven able men and prepares to accompany Pim back to the *Resolute*. There, he will confer with Captain Kellett in person. Before he leaves, he orders Cresswell, Piers, Wynniatt, and Miertsching to follow in one week's time with two dozen of those men in most critical need. He further insists that all possessions be left aboard the *Investigator*.

"Think of nothing," McClure says, "but bringing those sick and stricken men alive to Dealy Island."

As their guests prepare to take their leave, they find themselves amid the morning distribution of rations. The meal portioned out consists of a weak cup of sugarless cocoa and a few bites of bread. Witnessing the pathetic scene, Pim is overcome. He gathers himself, rushes up the ladder, and out to his waiting sledge. There, he retrieves a slab of bacon, marches below, and ceremoniously places it on the table before the officers of the *Investigator*. It is a breakfast they will never forget.

The body of Able Seaman John Boyle is prepared by his messmates. It is wrapped in black cloth, carried up the ladder, and laid in a freshly built box. The casket is draped in an English flag, carried down the gangplank, and placed on a sledge as the ship's bell tolls and the ensign is lowered to half-mast. The sledge is drawn by eight sailors toward an open grave hacked out of the rubble on shore. They are preceded by the ship's marines, marching in military order, followed by the rest of the company in solemn procession. The captain's conspicuous absence is somewhat relieved by the belief that help is on the way.

Armstrong is moved by the touching scene. As the cortège advances, he feels a sense of anguish in the face of a death that could have easily been prevented. Circumstance has placed in him in an untenable position, a position he finds painful in the extreme. These men suffer primarily from the lack of the one medicine—food—that is not within his power to give them. Even the dose of lime juice has been reduced to pitiful, trace amounts. It is not nearly enough to prevent or cure—it can only delay. Too often a doctor does not know what ails his patient and is therefore unable to provide comfort or alleviate pain. It seems far worse to know the enemy precisely, only to be disarmed in the heat of battle.

The procession comes to a halt on a rise above the beach. At graveside, the English Burial Service is read. With the words *ashes to ashes, dust to dust,* the coffin is lowered into solid ground. The bite of shovels and the thud of hard clods fills the air until the casket is covered and the pile tamped down. The marines take their places, shoulder their rifles, and fire three volleys over the grave and across the ice. At last, they lead the whole procession back to the ship with St. George's Cross raised high and flying.

Three days later, the ship's company is reduced again. Able Seaman John Ames's heart succumbs to dropsy (edema) brought on by starvation and scurvy. Another sailor lies close to death. When Gunner's Mate John Kerr was first brought to the surgeon, he was unable to stand on his own. Now he is so incapacitated he is unable to even raise his arms. Suffering from the worst effects of scurvy, he is now bleeding from his mouth and rectum. Miertsching visits the sailor, enquiring about his state of readiness for the life everlasting. Kerr appears calm, resigned to his fate. He tells the brother that he believes he will soon start a better life, free of sorrow and pain. The next day, Kerr follows Ames and Boyle.

Members of the first party to be evacuated to Dealy Island are allowed only two pairs of socks in addition to the clothes they wear. Five of those men slated to travel still lie in sickbay. Miertsching tries again, but is unable to make Wynniatt understand what is about to happen. The rest of those chosen to go are anxious to be on their way to the *Resolute*, leaving behind hunger, death, and the sight of Mercy Bay.

The brother seals and addresses his personal effects: four leather valises, a chest each of geological specimens, biological specimens, and Esquimaux weapons. When this is done, he grips his journals, feeling bitterness and grief at the thought of being forced to part with them. He would gladly trade all he owns for these simple notes. Finally, these too he surrenders.

The brother returns to his empty cabin. He closes himself in one last time and then lies down to wait.

19.

EXODUS

The weather does not encourage. A snowstorm has blown in and visibility is diminished, but the sledges are packed and everything is ready for departure. As Miertsching attempts to shake off the vestiges of sleep, a wave of nostalgia overtakes him. He counts the blessings he has received within this sturdy hull. Here he has spent both lonely and happy days, learning much about himself and the wide-ranging nature of men. The *Investigator* has carried them through numerous calamities, sheltering and protecting them from the harshest conditions imaginable. The time has come to leave her behind—not to begin an impossible march south and east across the continent, but north to salvation across the last link of their Northwest Passage.

Despite the appearance of Lieutenant Pim and the abandonment of McClure's plans, the men selected to remain onboard are kept on short rations. On this day, however, an extra meal is prepared and the entire crew dines together at noon on the lower deck. Those soon to depart are unable to contain their nervous anticipation, while the others listen with a mixture of envy and sadness. Then all men stand and exchange heartfelt wishes for safe travels and fortitude.

Following the meal, sailors begin appearing at Miertsching's door. They come to say goodbye and to wish him well. When his small cabin fills, he offers up a prayer that seems to leave a deep impression. When these men depart, still more arrive to say farewell and to thank him for the friendship he has shown. The brother is especially touched when one young sailor knocks on his door. When he first met this coarse and untamed person, he judged him a brazen sinner. Over the past three years, however, the sailor has learned to read and write as well as develop a fear of God. The brother counts his reformation a personal success. The sailor thanks the brother for his kindness and attention. In return, he gives him a collection of sixteen songs and hymns of his own composition as a souvenir of what he leaves behind.

And then the lower decks suddenly clear. The thump and shuffle of men up top signal that the time has arrived. The brother steps down the corridor, up the ladder, and into the light above. Gathered there are the officers and crew chosen to remain behind. The brother takes his place opposite them, alongside Cresswell, Wynniatt, and Piers. He stands at attention as Able Seaman James Nelson, the ship's uncrowned poet laureate, reads a composition penned especially for the occasion.

At last, my lads, we're about to part,
Some for our native shore,
And after changing years, perhaps,
We part to meet no more . . .

When you depart, dangers may oft
Beset our chequered way,
And troubles oftentimes arise,
Remember this and say:
I'll put my trust in Him above
Who calms the troubled sea;
And that bright Eye that's still aloft
Will watch over me.

No man remains unchanged by what has transpired here. These men, who once gave the brother such great and frequent offence, now show their care for him. Reduced as they are in body and might, they have measurably grown in spirit. For his part, the brother has learned to look beyond rough exteriors to what light may shine within. This is something he instinctively does with his beloved Esquimaux. He has begun to extend some of this grace to his fellow Europeans.

Down on the ice, the brother surveys the loaded sledges and then slips into a harness. The remaining men line the deck and offer up three cheers. Lieutenant Cresswell takes his place at the front of the column and signals the start of the march. He is followed by a half-dozen of the weakest men, charged only with conveying their own frail bodies across the floe. They are followed by the sledge teams and members of the relief party, which includes the ship's surgeon. Armstrong helps pull the load several miles, until the blowing snow forces his return. He watches as his individual shipmates form a contiguous black line, inching its way north through hummocks and drifts. He watches until the line disappears into billowing folds of white.

—

The brother awakens to a temperature of minus-25 degrees. After seven mostly sleepless hours inside the tent, he and his comrades rise, eat, and continue their sluggish march. The loads are not especially heavy, and had the men been sound, their task would not be so extreme. But by the second day, those unable to pull now cling to the sledges for support, further increasing the load. Several of the harnessed men find themselves suffering under the strain. When large pressure ridges and heaves must be crossed, the sledge teams are reduced to crawling on hands and knees.

The coming week brings more snow, then fog. At this latitude, their compass is an unreliable guide. As the sun, stars, and shore are all lost from view, Cresswell must resort to other methods of reckoning. At one point, the party catches a glimpse of a ptarmigan, which are known to cross between Banks and Melville islands. Intuition aside, the flight path of this bird is their most reliable guide.

The party now sleeps twice during each twenty-four hour day, a repeated cycle of five hours in the tent followed by seven hours' travel. Each time the lame stagger to a new campsite, they stand around and helplessly watch Miertsching and the few other able-bodied men set up the tents. Unable to help in any way, they are barely able to keep from keeling over. Of the nine men who eventually squeeze inside the brother's tent, five are so incapacitated that they must be wrapped up and helped into their blanket bags.

On day nine of their trek, at three o'clock in the morning,

the party arrives near the shore of Melville Island. After the brother helps erect the tent and puts the weak to bed, he ventures out on his own.

Ignoring his deep exhaustion, he climbs a low hill in the soft twilight and looks back out over the crossing they have made. He feels a swelling pride. A humble Saxon now stands on the shore of that Arctic island made famous by Sir William Edward Parry, in full view of the passage sought by mariners for over three hundred years. Ice-choked, treacherous, and useless for its intended purpose, it is the Northwest Passage nonetheless, and Johann August Miertsching can claim a small part of its discovery as his own. Now all he wants is to leave it behind. Distracted by hunger and cold, he hikes back down to camp for a meal of pork and biscuit, washed down by snowmelt and rum.

The weather improves, but the health of the men declines. Two are now so lame they must ride atop the sledge. Even with a reduced pace, many of those who cling to the sides are barely able to stagger along. The lunatics, Bradbury and Wynniatt, manage to make matters worse with their continued nonsense and raving.

On the first of May, at 7 a.m., Cresswell pulls out his cold brass telescope and holds it to his eye. He focuses on a dark spot to the north that, with any luck, is Dealy Island. A few hours later, he stops to focus again. He hands the telescope to Miertsching, who sees what appear to be the masts of two ships about fifteen miles away. At their current pace, they might be reached by nightfall tomorrow. A wave of relief passes through the men as they sense the end of their trial draw near. Both the able and the weak tap a previously unknown store of strength and press on with renewed purpose.

The next day, breakfast is followed by what each man hopes will be the final prayers said huddled on the floor of a tent. They pack the sledges and then set out in the hope that relief will be met following one, final push.

As they approach the frozen ships, the Investigators notice a flag rise up and unfurl. Men are seen coming off the vessels, marching in their direction. Soon, they are met by captains McClure and Kellett, along with a party of officers and crew. Seeing the pitiful condition of the refugees staggering in their direction, the men of the *Resolute* and *Intrepid* rush to take up their harnesses and shoulder the walking weak. The Investigators are moved by their warmth and kindness.

Following a quick triage, the invalids are sent to the *Intrepid*, ready and waiting as a hospital ship. Only six of the twenty-eight arriving Investigators are deemed well enough to join the *Resolute*. After so many months in each other's company, the rent feels disorienting, abrupt.

Captain Kellett conducts Miertsching to his own cabin, where he is able to wash for the first time in sixteen days. He steps out of his ragged clothes and into fresh underwear and outer garments provided by Kellett himself. This is followed by a breakfast with both captains, where a steaming cup of coffee is placed before him—the first in two and a half years. After this, he lies down in new quarters, where he sleeps until awoken to eat again.

As his officers and crew are made as comfortable as possible, McClure is summoned to a series of meetings with Captain Henry Kellett, a grey-haired fellow Irishman known for his disarming charm. After distinguishing himself in the Opium War, Kellett achieved the rank of post captain and is McClure's senior officer. He is shocked by the condition of the castaways now under his care. They have arrived in a wretched state—and with the report of two more deaths in Mercy Bay. Where before he may have been inclined to believe that McClure was capable of returning to the *Investigator* and rallying his remaining men, he is now circumspect. McClure disregarded his advice to await his consort, the *Enterprise*, when he entered the ice nearly three years before. And yet this audacity earned him the discovery of the Northwest Passage. A similar impetuous act now, Kellett believes, would needlessly endanger the lives of men who have already suffered beyond duty's call. But McClure won't easily stand down. Money, fame, honour, and history all hang in the balance. Proposals and counter-proposals are made until at last Kellett is forced to wield authority in a way he hopes the Admiralty will not view as obstructing a fellow officer's initiative.

McClure will be allowed to return to the *Investigator*. However, he will be accompanied by the *Resolute*'s chief surgeon, William Domville—an unprejudiced observer. Domville, together with the *Investigator*'s own surgeon, will conduct a thorough examination to determine the fitness of the officers and crew at Mercy Bay. Finally, the men themselves will not be compelled to stay. The decision whether or not to abandon ship will be their individual prerogative. However, should twenty able-bodied men volunteer to remain and await the breakup of the ice, McClure will be permitted one last attempt to sail his ship across his newly discovered Passage.

Forty-eight hours after arriving at the *Resolute,* all six "able-bodied" Investigators complain of severe pain in their limbs and are now mostly confined to bed. The ache in Miertsching's right arm and leg is such that he is no longer able to dress himself. Despite the pain, he hobbles down to McClure's quarters, where he finds the captain packing his kit. Kellett's compromise is explained; McClure and Domville will set out this evening in the still-favourable light.

"I know you worry about your journal," McClure says. "It is in safekeeping aboard the *Investigator.* Should I be successful, I will send it back with Dr. Domville when he returns. If the ship must be abandoned, I will bring it to you myself."

The brother is grateful for this consideration and promise. As this may well be the last time they see one another, he wishes his friend good luck—and the realization of his long-held dream.

By his third day aboard the *Resolute,* Miertsching has become so lame he can barely haul himself down the corridor. Regardless, he makes his way aft to Kellett's cabins, where he learns of the captain's plans to evacuate fourteen of the sickest men to the *North Star,* two hundred miles to the east at Cape Riley, and from there on to England. They will be led by Lieutenant Cresswell, whose impatient and vociferous father awaits his son's return.

The way of escape now seems so suddenly close that the brother need only step in the right direction. He is not a navy man, nor an Englishman, but a paid volunteer who has endured much for the British Crown. He can gain no possible advantage or advancement by remaining on the shores of the Polar Sea. Surely, he has served far beyond any expectation. God's work beckons—along with a few of the comforts and joys of civilization. The moment has come to speak up for himself, to act on his own behalf.

"Sir, I too am now lame. Unwell as I am, I would gladly be part of this party."

"I am sorry," Kellett replies, "but I cannot grant your request. I may very well require an interpreter next summer among the Esquimaux of Baffin Bay."

At this, Miertsching humbly bows and retreats. Self-denial and obedience are pillars of his religion and philosophy. True strength, he reminds himself, is revealed in the act of submission.

Back in his new quarters, the brother hastily writes two letters home, informing family and friends of his continued existence. These he places in Cresswell's hands, along with the wish for Godspeed across the ice and sea. He then turns to Wynniatt and offers a heartfelt goodbye. But the baffled mate persists in his dementia. He has little sense of what's past or what lies ahead. Perhaps, over the course of this next march, the knot within will come undone and he will find himself again. Failing that, maybe relief will come when he sails past that final floe and puts the ice forever behind him.

20.

SHOW OF HANDS

Spectres haunt the ship. The pale, haggard men now stalking the decks appear to Armstrong like some kind of living dead. The invalids sent away last month had the benefit of eight weeks on improved rations before their departure. The officers and crew who remain have survived on dwindling portions for over a year and a half. The changes wrought on their bodies are plainly revealed in the light of spring, each crag and shadow on their faces defined. As ship's surgeon, he was well aware of their prime physical condition before they left England, before the cold, sickness, and starvation settled in. Now, the very sight of them is hard to endure.

On May 19, six weeks after he left the *Investigator*, Captain McClure is seen approaching from the north. He arrives with the surgeon Domville and orders from Kellett that are made immediately clear. What was not made clear to the captain of the *Resolute* was the true condition of the men aboard the *Investigator*. Domville had been asked to give a report on McClure's invalids as soon as they arrived. In addition to their wasted physical state, it was noted that these men were possessed of a vacant stare and that their intellectual capacities had been diminished. For Armstrong, it is another sad vindication. Not only has Kellett called for the unbiased opinion of his own surgeon, but also that of himself, the ranking medical officer aboard the *Investigator*. If Kellett is seeking a complete and unclouded view of the true state of these men, he will get it in vivid detail.

If McClure is chastened by these orders, he is determined to keep it contained. He only wants these examinations to take place as quickly as possible. The two surgeons acquaint themselves with one another, then set about examining the thirty-five men remaining aboard the vessel. They are faced with a parade of spotty thighs, painful joints, bleeding gums, exposed ribs. The surgeons carefully document the various maladies they encounter along with recommendations. Agreed in their assessment, they send for McClure.

The men aboard the *Investigator* are unfit to face another winter. It is a refrain he has been hearing for some time, only now in harmony. Without exception, there exists in all of these men evidence of scurvy. Consistent with starvation, they also exhibit significant loss of muscle and strength. The surgeons concur, and will report that no man is fit for service. It is a reality that can be plainly seen even to the untrained eye. It is a truth that can no longer be contained or denied.

But the choice must still be put to the men. McClure calls an assembly of everyone onboard. He lays out the directive from Kellett, and his own wishful plan of sailing away and through the passage this summer. Twenty men are needed. When he calls for volunteers, his four officers—including

Armstrong—step forward. But it is a hollow gesture. Everyone knows that the result of the medical exam was a foregone conclusion. It is but an example to the men, as a show of honour and respect for themselves and, perhaps, McClure. Only four members of the crew follow their lead. The charade is finished.

McClure, so sure of himself and his own abilities, is stunned and humiliated at the result. He dismisses the men and retreats to his cabin.

When the captain at last appears, he orders the ship's company to muster on the quarterdeck. There, he tells the men that, after all they have accomplished and endured together, he is sadly obliged to order the *Investigator* abandoned. All that remains is to land and secure the remaining boats, provisions, and stores for Sir John Franklin or Captain Collinson, should he and the *Enterprise* still be somewhere in this vicinity of the Polar Sea, or for any other traveller who, in years to come, finds himself in need at Mercy Bay.

The announcement is met with profound relief.

"Further," he says, "I order everyone placed on the full allowance of provisions."

And thus ends their days of privation, hunger, and suffering—or so Armstrong is determined to believe. He immediately asks the captain to allow the men as much lime juice, preserved vegetables, and other food as they desire. He would limit nothing but alcohol. This request is granted. Having gone twenty months without a decent meal, the men devour their food with passion. The effect on their startled digestive systems is immediate and pronounced. Painful stomach aches are followed by lethargy and loss of appetite. Some even gorge their way to sickbay.

His work now concluded, Domville's small party sets out for the *Resolute* with the results of the medical assessment. Armstrong sends Sainsbury with them. The first mate was too weak for the previous evacuation, but has rallied somewhat since. In his fragile state, his best chance is to travel with this smaller group, under Domville's undivided attention and care.

The process of gutting the ship begins immediately. A work party is formed and a depot established at a safe location above the beach. Over the course of the following week, the bulk of their stores and provisions are hoisted out of the holds, over the deck, and across the ice to shore. Anything that can provide relief is off-loaded. It is just such a pile of goods that could have relieved their own misery over the past two years:

600 lbs. salt beef	1,000 lbs. sugar
1,600 lbs. salt pork	435 lbs. chocolate
3,000 lbs. preserved meat	126 lbs. tea
6,420 lbs. flour	484 lbs. tobacco
1,000 lbs. biscuit	26 gal. rum
112 lbs. suet	20 gal. brandy

Added to this is a year's worth of clothing for thirty men, plus coal, a boat, spars, rope, powder, shot, and arms. This task they perform without bitterness or complaint. Observing the withered crew stockpile these riches for other castaways, Armstrong is moved by their sense of duty and honour.

Once most of those useful things are safe and secure on shore, a cairn is erected on the neighbouring hill to alert weary travellers. Inside is placed a record of the events that transpired here and the reasons for abandoning ship.

Their last task is a painful one. A tablet is carved and placed above the three graves on the beach. It contains an epitaph to the memory of the men who lie here, declaring that they fell nobly, in patriotic duty. The act of placing this modest marker has the feel of a solemn rite. Taking in the lonesome, mournful scene, the surgeon can't help but feel the loss and shame that these men weren't preserved long enough to learn that salvation was near.

And then that longed-for day arrives. As with the last sledge party to leave the ship, the weather threatens to hem them in. Low grey cloud sucks the light from the sky and creates an atmosphere of foreboding. Snow begins to fall.

On June 3, the *Investigator* stands doubly anchored and made ready for their departure. Everything is put in perfect order, snug in its place, ready for anyone who might one day need shelter in such a remote and useless place. And then she is cleaned from stem to stern. It is an act of respect to these imagined guests, themselves, and the body of the ship. Finally, cabin doors are closed and, one by one, the men climb the ladder and assemble above.

At 5:30 p.m. the crew awaits orders and some sense of resolution as McClure, Armstrong, and Haswell make their final inspection of the ship. The grim-faced captain then addresses his men with a few curt words and orders them down to the ice. The ensign is hoisted at the end of the hull, the pennant set atop the mainmast. Both take to the wind as the officers step over the side.

The carpenters are the last to go. They batten down the hatches and then pull the gangway free. The men now stand together on the ice, looking back with the realization that here they will leave a part of themselves behind. For all good mariners, ships are living things that require constant attention and care. And this one returned their affections well. She was their home for three and a half years, but she was also a prison and would surely have become a tomb. The time has come to turn their backs and consign her to her fate.

The four sledges weigh between 1,200 and 1,400 pounds. Despite the increase in nutrition over the past twelve days, the men remain thin and weak. The privileges of rank are set aside as Armstrong and his fellow officers take up a harness with the crew. At 6:10 p.m., they lean as one and take their first steps in silence.

Over the next few days, the skies clear, the sun comes out, and some of the men go snowblind. They march on, blindfolded—slipping, falling, tangling themselves and their mates in the ropes. Some grow so tired they are no longer able to help; others are eventually strapped atop a sledge.

Everyone suffers from thirst. Snow is eaten continuously. Then, to avoid further inflaming their mouths and throats, the men squeeze snowballs in their hands and suck out the resulting melt. As temperatures rise and the surface starts to thaw, icicles form on the hummocks and blocks. These are broken off and carried in pockets to be sucked when needed. At night, the temperature plunges again, freezing wet boots to numb feet.

The ice. The cursed ice. Their long and cherished dream was to see it rapidly melt away. Now, it seems cruelly determined to grant this wish only to prevent their escape.

Twelve days after abandoning ship, the party camps off Winter Harbour. Now the thaw comes on with such resolve that it becomes hard to find a bed of snow in which to pitch their tents. When one is found, it isn't long until those lying inside find themselves soaking in a puddle. During the day, they splash through endless pools, some of which reach halfway up their thighs. Their sodden legs and feet become frostbitten when the temperature falls at night. Come morning, they find themselves crashing through fresh plates of ice, the sharp edges and shards slicing their canvas boots.

Armstrong is faced with the fact that the health of the men is fading. Their feet and legs are swollen. They suffer increased pain and stiffness in their joints and severe chest spasms from the constant agitation of the harness. And then—scurvy-ridden and nearly starved—an exhaustion sets in that will not be relieved.

Day fourteen. Camp is made within sight of Dealy Island and what appears to be the faint outline of ships. The Investigators take four hours' rest, then rise and gather round one of the countless pools. They scrub and splash their blackened faces, attempting to make themselves presentable for their old shipmates and the new acquaintances they expect await them.

They then pull toward that distant sight from which unblindfolded men are unable to tear their eyes. Left, right, fore, and aft, Armstrong hears men talk of nothing but the pleasures and comforts ahead, not least of which is the chance to see new faces, hear unfamiliar voices, and become reacquainted with the ways and cares of civilized men.

Miertsching has been watching from the deck of the *Resolute* since the party was first spied in the telescope at eight o'clock in the morning. The caravan appears to move at the pace of a winter's night. At one o'clock, he finally steps off the ship and marches in their direction. As the shadows coming toward him take shape and become recognizable forms, his heart sinks. The image of his shipmates emerging from the melting sea is one he vows never to forget.

Two lame men are lashed atop each of the four sledges. Others, unable to walk on their own, are supported by men slightly better off than themselves. Many hang onto sledges pulled by men who march with such a tottering gait that one falls every few minutes and is unable to rise unaided. And these were McClure's "able-bodied" men, destined to either sail the ship through a torrent of ice or face another winter trapped at Mercy Bay. This spectacle of misery puts the brother in mind of the doomed Franklin expedition, now eight years gone. Could they have possibly endured all this and more? The arrival of Pim, just before McClure's plan could be set in motion, is nothing short of a miracle. Had not God himself intervened, the brother believes, every man who set sail from England aboard the *Investigator* would have surely met his death on the Polar Sea.

Two miles out, the sledge party is first met by Miertsching, Kellett, and officers from both the *Resolute* and

Intrepid, who extend a warm welcome—along with armfuls of refreshments. Moments later, they recognize their old shipmates running, hobbling, shuffling toward them. Although their separation has been only eight weeks in duration, they seize their arriving comrades as if years have intervened. Greetings, news, and laughter fill the air. They take over the harnesses from their weary mates and pull the loads with apparent ease.

Amid the joyous reunion, Armstrong is struck by the startling improvement in the health and demeanour of the invalids he sent off only two months before. Some are so utterly changed he does not recognize them at first. Most have gained considerable weight and have the bulk and strength of sailors once more. The gaunt and hollow faces have filled out, and are transformed with radiant smiles at witnessing their shipmates' deliverance.

The party continues to grow as it reaches the ships. Armstrong looks for familiar faces in the crowd, then glances up to see that even the ships themselves are brightened with pennants and flags in honour of their arrival. Still more men appear. They gather to receive the Investigators with three hearty cheers. In this moment, Armstrong feels the realization of all their hopes and dreams.

Following triage, the surgeon, most of the officers, and twenty-two members of the crew are all sent to hospital aboard the *Intrepid*. There, they find that every imaginable preparation has been made, including the construction of temporary cabins. Armstrong and his comrades are received with humbling kindness. A luxuriant feast has been prepared, the likes of which they have not seen in years. The Investigators do justice to the meal with toasts to their generous hosts, each other, and the end of want and misery.

Aboard the *Resolute*, Miertsching learns that the *Investigator* was abandoned after most useful provisions were moved ashore. Trifles and ballast, such as natural history collections, were left sealed onboard. And then McClure explains.

"I am sorry, but I was unable to bring your journals."

The brother estimates they would have weighed no more than two pounds. The captain had given him his word.

"If I was to make this exception, I would have been obliged to also bring the journals and records of my officers. And this, I could not do."

But a Moravian missionary is not an officer of the British Navy. Surely an exception could have been made, especially for a civilian, a man of God who has given up valuable years at his mission. The captain cannot misunderstand what these journals mean to him. Miertsching was under the impression that McClure was more than his senior officer. So much conversation and council has passed between them. Does he not trust him with the story of their voyage? He feels the sting of disappointment, and yet he instinctively forgives. The men who arrived today were little more than skin and bones. They could and should have thought only of saving their own fragile lives. Surely, this is no time for thoughts of a petty, selfish kind.

"Mr. Miertsching, please. Borrow my journal. Use it as a guide to revive and reconstruct your own memories."

McClure hands over his personal journal, along with some writing paper—which aboard these ships is scarce in the extreme. To this, Captain Kellett adds another dozen sheets of paper. The surgeon Domville offers two pens, another officer donates some ink. Beyond what the meek shall inherit tomorrow, humility always—eventually—sees the brother through each day.

But time is of the essence. Breakup is just weeks away. He must be diligent, get it all down, capture the memories so vivid in his heart and imagination. This he must do before the ship sails the high seas come August or September—before their imminent voyage home.

21.

UNCOMMON PRAYER

Candlelight through sailcloth casts the faintest glow. It is enough, with time, for Miertsching's eyes to adjust and allow him to get his bearings. He lies atop a makeshift berth, surrounded with the canvas walls of his new lodgings aboard the *Intrepid*. It is but a tent inside a ship, tacked up to make do during a speedy summer run across the Atlantic. Transferred from the *Resolute* a few months ago, the brother came with only the clothes on his back, his paper and pens, and two woollen blankets, one of which came with him all the way from the *Investigator*. The blankets, like his beard, are now covered with frosted breath. They provide little protection from the tempterature of the lower decks, which never climbs above freezing. The canvas walls cannot contain even the warmth generated by his own body. He listens to the coughs and murmurs of his neighbours for a time, then sits up and dons his coat and boots. He steps through the flap, up the hatch, out into the start of what will most assuredly be a long and mournful day.

November 16, 1853. One hundred and seventy-five miles east of Dealy Island, the ships lie beset. The *Intrepid* is a 400-ton, 150-foot, 60-horsepower screw steam tender. Francis Leopold McClintock, commander. Four hundred and thirty-two paces across the ice lies Kellett's *Resolute,* a three-masted barque, almost the exact same dimensions as the abandoned *Investigator.* Together, the ships house 175 men, sixty-three more than for which they are provisioned and equipped. All parties have long since returned from exhaustive searches for Franklin and his men, over the course of which not a single trace was found. Both ships now suffer from a desperate lack of space. Even Captains Kellett and McClure must share a single cabin.

For three days now, the ice has been quiet and calm. A series of sightings were taken, confirming the fact that both ships are now fixed at latitude 70 degrees, 41 minutes, 35 seconds North, longitude 101 degrees, 22 minutes West—about twenty-eight miles to the southwest of Cape Cockburn on Bathurst Island, in the vicinity of the North Magnetic Pole. Had the weather been clear today, the shore would be visible from the crow's nest. Much too far for hunting or recreation, it is all but lost to them.

Rescue started out well enough for the Investigators. When they first arrived at Dealy Island, their health rapidly improved. Mindful that their presence aboard these ships was a tax on others' provisions, they were keen to contribute to the common good. Some were soon able to join hunting parties to Melville Island, where game was plentiful and their efforts met with success. Over the course of a single week, Armstrong and Pim together brought in a muskox, nine caribou, ten hares, eighteen geese, thirteen ducks, and sixteen ptarmigan. Upwards of ten thousand pounds of game was eventually secured during the summer season, the yards and rigging of both ships festooned with fresh meat set out to freeze.

Then, on August 10, the sea came welling up through rents in the ice, scrambling over top and running across the surface as far as the eye could reach. The crews of both ships had been ready, impatient to sail, and saw in this flood hope for the sudden breakup of the ice. But following this event, the sea and ice again grew still. Parties sent out to search for signs came back to report no significant change. The ice was holding firm everywhere they went—even along the shore.

Within a week, northwest winds rose into a gale strong enough to make the ice undulate. This storm coincided with a high tide, which eventually set the ice in motion—smashing the *Resolute*'s rudder to pieces. Then ice, ships, and sea moved as one, and the men soon found themselves thirty-five miles from Dealy Island. To ready themselves for that moment open water should appear, the splintered rudder was removed and a new one hung in its stead.

The balance of the month was spent drifting so far from land that none could be seen. The brother marked his thirty-fifth birthday in quiet contemplation, weak and weary of spirit, willing the ice to part and make way. If only he could have convinced Kellett to let him go with Cresswell and Wynniatt to the *North Star*, he would have been free of the perpetual unknown, the maddening sense of liberation one moment, the noose tightening the next. He would be back in the land God created as the proper, fertile home for both man and beast—not entrapped in sterile ice like an ancient amber fly.

By September 3, Kellett had made it clear that all plans of continuing the search for Franklin in Baffin Bay were now abandoned. His only hope was to make it through the ice and back to England along the most direct possible route.

A week later, the ships were sailing east, and soon put twenty-five miles behind them. Eventually, the *Intrepid* took the *Resolute* in tow and—under both sail and steam—set a course east through opening leads, to the thrill of every soul onboard. With each mile made in the direction home, the men grew in confidence. So much so that, by September 10, they were laying bets on the exact day they would step on English soil.

On September 11, they arose to a sight that took their breath away: ice pressing in against the ships' hulls from every point on the horizon. Nowhere could a pool of open water be seen. The severe cold, ushered in by the floes, took hold and bore down. Hour by hour, the ice muscled in tighter and became more firmly set. The limits of their ingenuity and strength would move the ships less than a dozen feet. They found themselves facing a well-known opponent, one that still managed to surprise and deceive. They should have known better than to drop their defences and expose their hope again in such a naked way. Within two days, the new ice had grown five inches thick.

The ghost of Franklin had not failed to take its pound of flesh. Before being swept away from Dealy Island, Kellett—confident of a quick escape from the Arctic—established a depot for those lost men, containing a good portion of their remaining food and clothing. He kept only nine months' worth of provisions. On September 18, observing their fate circling in, he put the crews of both ships on two-thirds rations. The Investigators, having survived far worse, shrugged off this news. What was nearly impossible to accept was the idea that yet another Arctic winter must now be endured.

The cost of this realization was far too much for First Mate Hubert Sainsbury. He had been hanging on in the hopes of seeing his homeland again. Like the world in which they found themselves trapped, his fair young face had drained of any colour.

His own health much improved, Miertsching took to proselytizing again—with decidedly mixed results. Some men have no patience for religious talk of any kind; others suspect the brother of attempting to draw them away from the Church of England. Undaunted, he continued visiting the sick. Daily, he sat and spoke with Sainsbury, with whom he had been messmates since the *Investigator* sailed from Plymouth. When the first mate's powers of speech began to fail, the brother would read him scripture and holy tracts and visit both day and night. Then, two days ago, Sainsbury suddenly called out for the captains and officers. One after another, they arrived to pay their respects. He greeted each in turn and, with clasped hands and whispered words, asked their forgiveness for his numerous mistakes and shortcomings. He made his peace with them. The brother was left to hope he had made a similar overture toward his creator.

Weakened by starvation and scurvy, Sainsbury had for many months been suffering the fever, night sweats, wasting, and blood-tinged cough of consumption (tuberculosis). In the end, the only medicine that seemed to offer hope was the belief that the ice would break and they would soon be sailing home. When it became clear that they were doomed to remain another winter, the disease overcame his last defences. Armstrong had hoped that if they could reach civilization, Sainsbury might live to see his twenty-seventh birthday. At half past midnight on November 14—eleven days past the setting of the sun—his suffering came to an end.

It is 57 degrees below zero when the ceremony begins. On the deck of the *Resolute,* in the midday gloaming, Captain Kellett reads aloud the first words from the Book of Common Prayer customary for burial at sea. The body, wrapped in a sailcloth shroud, is then hoisted up and carried down to the waiting sledge. There, it is covered in the Union Jack as the ship's ensign, lowered to half-mast, rails against the cold north-west wind. The sledge is drawn by six petty officers of the *Investigator,* and is followed by the crews of all three ships. The procession moves two hundred paces across the ice to a smooth, flat stretch where a square hole has been hacked through to liquid sea. The sledge is drawn into position as the men gather round. And then the balance of the service is read, the words curling out on vapour clouds about Kellett's lips and cheeks.

"We therefore commit his body to the deep, to be turned into corruption, looking for the resurrection of the body, when the sea shall give up her dead, and the life of the world to come, through our Lord Jesus Christ; who at his coming shall change our vile body, that it may be like his glorious body, according to the mighty working, whereby he is able to subdue all things to himself."

At that word *deep,* the sledge is carefully tipped and—like a weighted cocoon—the body glides through the slush that has begun to form. Without a sound, it is engulfed and pulled down beneath the wind, their gaze, their boots, their ships' immobile keels. After the briefest ripple, all is perfectly still.

To himself, the brother recites the more elegant conclusion from the Liturgy of the United Brethren, appropriate to the occasion.

Now to the sea let these remains
In hope committed be,
Until the body, chang'd obtains
Blest immortality.

The Investigators, now further reduced, divide again and return to one or the other crowded ship to face a fourth winter on the edge of the Polar Sea. Within the hour, a fresh scar has formed over the grave and the ice is healed again.

22.

BETWEEN SHIPS

Petruchio is having none of Katherina's cheek. In fact, he is intent on showing her a woman's proper place—bent to his will. In this case, the shrew being tamed is a member of the *Resolute*'s crew made fetching with a frock, provocative moves, and flattering light amid the props and stage built on deck beneath the canvas housing. The carpenters worked for two solid weeks. The cast, composed of sailors, spent many hours more out on the ice, beyond earshot, running their lines in private. The launch of this elaborate production, postponed by the death and funeral of First Mate Hubert Sainsbury, is now sailing through both its opening and closing night.

Accompanying the immortal words of William Shakespeare is a newly minted orchestra consisting of a drum, a home-made fiddle, and six fifes fashioned from a copper curtain rod. Programs are printed on red silk for the captains, blue silk for the officers, paper for the crew. The audience, fortified with punch (officers) or a pint of beer (crew), clap enthusiastically in appreciation and to keep their hands from freezing. The temperature is thirty below, despite stoves placed at the corners of the stage that serve as little more than decoration. The audience finds most everything about the production marvellous. Among them, however, sits one man who fervently believes that such talent and energy could be better spent on more edifying and uplifting pursuits.

Intermission consists of songs and a style of vaudeville by the *Investigator*'s own Charles Anderson, who, dressed in black dinner jacket, draws uproarious laughter and applause because he is a Negro attempting to pass himself off as a proper gentleman. While the set is changed, the audience blows over their fingertips and stamps tingling feet. Intermission is followed by the popular British farce "The Two Bonnycastles," this time played by the officers. Theatricals conclude at half past nine, but the drinking and merriment carry on to midnight. Diversions such as these, common on other Arctic discovery ships, are a revelation to the men of the *Investigator*.

Noting that much of the plot and substance of the plays seem lost on illiterate members of the crew, Kellett institutes a twice-weekly lecture series, to be given by the officers and delivered in simple, intelligible language. The subjects of these "time shortenings" range from the natural sciences and the systems of oceanic currents to mechanics, construction, and Arctic travel and exploration. Miertsching finds these diversions much more to his taste. However, variety shows—featuring singing, burlesque, and masquerade—also find their way back on stage. Of these, the brother finds himself drawn to the magic tricks and shadow plays—as long as they do not conjure images of an unwholesome character.

In good weather, social interaction between the *Resolute* and *Intrepid* offers regular relief. Toward the end of the year, when temperatures plunge beyond the point of endurance, a telegraph wire is strung between the ships with galvanic batteries at either end. This "blitzpost" becomes the source of endless jokes and amusement, including games of chess played between opposing crews. With so much time on his hands, Miertsching begins offering dogma and moral correction aboard the *Intrepid*. His pious and disapproving humour begins to grate on his new shipmates, who mock him with heathenish taunts and opinions.

Amid the usual uproar on New Year's Day, 1854, a member of the *Intrepid*'s own crew lies close to death just beyond Miertsching's canvas wall. The man, a thirty-four-year-old marine, is a committed freethinker. The brother is unrelenting in his attempts to convince this weakened man of his need to repent and turn from his convictions. The marine has no choice but to forbid the brother from approaching his bed for fear of having to pass his final hours undergoing yet more religious harangue. Nevertheless, the brother continues to pray for his conversion. The marine dies as he lived—self-possessed to the end. Observing this death, the fourth among the *Intrepid*'s forty-member crew, the brother is left to marvel at the fact that only four in sixty-six Investigators have died thus far, despite being in the Arctic for twice as long and having endured extremes of privation and misery. The brother prays that he may be ready that day, that hour, when he himself is called.

Winter aboard the *Intrepid* continues cold and cramped, but with adequate provisions and plenty of candlelight. On the approach of spring, Kellett dispatches two sledge parties from the *Resolute*. The first has orders to search Prince of Wales Strait for traces of the missing *Enterprise*, the second to cross over to Mercy Bay, survey the state of the *Investigator*, and recover the journals abandoned there. At last, Miertsching hopes, he will have access to his complete and unabridged memories.

In the afternoon of April 10, a sledge arrives from Cape Riley with orders from Sir Edward Belcher, commander of the Arctic Squadron. In his opinion, this expedition has come to an end. Kellett is to abandon both the *Resolute* and *Intrepid* and send everyone packing to the *North Star*, two hundred miles to the east. While immediate preparations are made for evacuation, Kellett dispatches a letter of protest to Belcher, asking to be allowed to remain with fifty men. If given another chance, he believes he can escape the ice and sail his ships home.

While Kellett awaits a reply, the first party of evacuees prepares to leave. Armstrong and Pim are to depart on April 11 with twenty-two men. They will be followed a day later by three sledges under the command of McClure, Piers, and Miertsching. Before their journey begins, Kellett calls an assembly aboard the *Resolute*. There, he publicly thanks the Investigators for their exemplary conduct, discipline, and character—qualities rarely seen in such abundance aboard ships at sea. Following this address, he hands McClure a letter, addressed to the Admiralty, echoing this praise.

The Investigators abandon ship once more. This time, they do so without fear or regret, filled instead with optimism and confidence that they will reach the end of their march in twenty days and, soon after, the open sea. The launch of this sledge journey is unlike those that have come before, for both the wisdom they bring and their faith in what's to come. This time, the ice does not hinder their every step, but presents a mostly clear and level path in the direction home. In only a few places do they encounter heaves that rise to forty-five feet, and these they do not find formidable. Everyone has plenty to eat, most men are healthy, and spirits run high. Only a single sailor is too weak to walk: Able Seaman Thomas Morgan, who has been ill for the past two years. He alone is forced to ride atop a sledge.

The Investigators are twice reminded that they trespass a foreign realm. One night, the head of a polar bear appears inside a tent and begins to sniff at the sleeping men packed tight in their blanket bags. One man awakens, draws his musket, and fires a shot. The ball misses its target and passes through the rope holding the canvas aloft, collapsing the tent atop the startled men and bear. The residents of the adjacent tents—jolted awake—quickly dispatch the intruder. A second visitation occurs again while the men are asleep. This time, the bear enters the brother's tent and begins inspecting the contents. One quick-witted sailor grabs his knife and slices a seam in the tent, and the men roll out to safety. They run to the sledge, load their muskets, and fire together in the style of a summary execution.

A kind of lightness and even joy distinguishes their days as they travel across the frozen sea. And rousing events such as these, which end with no man harmed, add to a sense of shared destiny.

Miertsching is amused with the invention of a "vapour cure" for night frost. Awakening each morning inside the tent, the men find condensation from their combined breath seems even worse than before. They often wake to find their long beards frozen firmly to their bedding. When this occurs, tentmates gather around the victim, pipes filled and lit, taking great lungfuls of smoke, brightening their glowing bowls. The combined output of six or eight pipes, directed at the subject's chin, brings him to the verge of suffocation but never fails to free him from bed.

The brother's marksmanship brings a surprise. One day, he spies a dark shape hunkered down in the snow. He grabs his rifle and fires a direct hit, but the target does not move. Assuming it must surely be dead, he approaches and finds that his ball has not struck a creature at all, but a tin chest containing fourteen pounds of coffee—part of a scattered depot left for Franklin and his men. His comrades send up a loud and enthusiastic cheer. From here on in, they will have their fill of coffee each time the kettle boils.

On the morning of April 28, after a march of fifteen days, the last Investigators reach their goal, the *North Star*—a twenty-six-gun frigate firmly beset in an unending sea of ice.

This latest vessel to which the Investigators are assigned lies but three hundred paces from Beechey Island, an insignificant speck in this broad archipelago of rock and ice. Miertsching visits twice. The first time, he strolls around its circumference in less than an hour. He then searches the tracts of bare, windswept ground and finds potsherds, barrel hoops, chips of glass, lengths of iron cable. These are not new discoveries. While he and his comrades were lost to the world, it was revealed that

Franklin and his men had passed their first winter here before moving on in pursuit of the Passage. In addition to the remains of a campsite, they also left a portion of their company behind. The brother locates the mounds, each marked by an oak slab commemorating the foreshortened life of an Englishman—thirty-two, twenty-three, and twenty years in length. The most recent marker bears the date April 3, 1846. Where they travelled after Beechey Island, no one seems to know.

Miertsching returns to the island once more. He and his fellow Investigators go to inter the body of one of their own, the fifth and—God willing—final fatality of their expedition. The body is laid to rest near Franklin's men, with a fresh board and an epitaph all its own.

SACRED
TO THE MEMORY OF
THOS. MORGAN,
SEAMAN OF THE H.M.S. "INVESTIGATOR"
WHO DEPARTED THIS LIFE
ON BOARD OF H.M.S. "NORTH STAR"
ON THE 22ND OF MAY, 1854,
AGED 36 YEARS.

Armstrong reflects on the sailor's sustained battle—a battle lost long ago to hunger and scurvy aboard the *Investigator*. As with First Mate Hubert Sainsbury, the news of a fourth year added to their Arctic sentence broke Morgan's will, and then the long sledge journey finally did him in. He was, in the surgeon's estimation, a most excellent man. The brother is mindful that, more than three thousand miles from this pile of stones, a woman and four small children eagerly await a husband's and father's return.

It is then made clear that Kellett's request to remain with his ships has been denied. Sir Edward Belcher reconfirmed his order to abandon the *Resolute* and *Intrepid* and evacuate all remaining men to the *North Star*. They arrive on May 28, a day after the latest shipment of cargo from the *Assistance* and the *Pioneer*—Belcher's own vessels—beset fifty-four miles to the north. These too he now seems ready to leave. One of the chests, marked NAUTICAL INSTRUMENTS, breaks open as it is being transferred to the hold. Inside are found souvenir pelts of Arctic fox and three dozen bottles of wine.

23.

DOMINION

October 9, 1854

Miertsching steadies himself over a basin and in front of a mirror. They say the ship is now barrelling along at up to eleven knots, but he is convinced the time has come to cut his beard. He has not put a razor to his cheeks in four years, and the growth has been extreme. After a time with scissors and blade, the brother's neck, jawline, the cleft of his chin emerge. It is a face both familiar and changed. These years have extracted a toll far beyond their normal fee. Regardless, he will step out of the wilderness and greet the civilized world with both a clean face and conscience.

The air has grown so warm that men are actually heard to complain. Since passing through the icebergs of Baffin Bay, crossing the Arctic Circle, and joining the North Atlantic Current, the Investigators have found themselves reborn into the familiar, temperate world once more. The heavy swell is now mostly past. The headaches and nausea, the weakness in the knees, recede. Here, warm water is pulled north from the equator toward the surface as cool water from the pole travels south and descends. Reliant on the push and pull of Arctic ice and cold, this current is in part responsible for the fertile, temperate climates of the British Isles and northwestern Europe. The brother would be surprised to learn that he has Arctic ice to thank for helping keep his world warm and green.

It was two months ago that the sea first cracked and lanes began to grow. Open water, however, remained some distance away. Everyone was obliged to join one last battle with ice as thick as twenty-five feet. With saws and powder, crews set to work cutting a canal twenty paces wide, nine hundred paces long. After three weeks of steady blasting and labour, the job was done and the *North Star* floated free of the shore fast ice. Three cheers flew from the deck at nine o'clock on the morning of August 26, as the sails were seen unfurled.

With the *Investigator, Pioneer, Assistance, Intrepid,* and *Resolute* all surrendered to the ice, 278 men were crowded aboard the one remaining ship. Among them, Miertsching and McClure stood together at the stern. They watched the floes fall behind and the expanse of sea open up to greet them. It was a sight neither man had seen in four years time—a view that moved McClure to tears.

Just as the *North Star* rounded Cape Riley, two of Her Majesty's ships were seen approaching from the east. The *Phoenix* and the *Talbot* were greeted with joy unleashed. They carried the first correspondence the Investigators had received since 1850, and the news from home brought both happiness and grief. After fresh supplies for Franklin and his men were stored at Beechey Island, Kellett and Belcher's crews were transferred to these vessels. The Investigators remained aboard the *North Star*, minus McClure, who was summoned to Belcher's side. He was forced to part company from the men with whom he had quarrelled, triumphed, and suffered for nearly half a decade. He bid farewell with sorrow.

It seems the brother has cleaned up just in time. The light at Beachy Head has been seen. This lighthouse is known to stand in front of the unseen cliffs of East Sussex, England, Great Britain, Europe—all now veiled in darkness. The lights of Hastings soon follow.

By eight o'clock in the morning, the *North Star* is anchored off Ramsgate awaiting a tug. When the brother joins the crowds on deck, he is torn between the sight of the land before him and the other ships at sea. Here, the maritime traffic is thick. He searches for a glimpse of the mission ship *Harmony*, which he knows will be ferrying brothers and sisters back from Labrador this time of year. He has looked for this little brig since Greenland's Cape Farewell. He continues his search, amid the great and endless stream of vessels, much to the amusement of the *North Star*'s captain and officers. They each bet a bottle of wine that the poor, homesick Saxon cannot possibly find the ship. He scans the sea all morning, resuming watch after lunch. And then, like a sign of God's good grace, the object of his search sails into view.

A trim vessel with white stripes approaches from behind. The brother grabs the telescope and, with a full heart, announces the approach of the *Harmony*. This is soon verified by those who disbelieved.

As the *Harmony* sails past, images of a polar bear and caribou appear painted in white on her stern. Among the crew on deck, three passengers are seen: a woman in European dress and two men wearing the native garb of Labrador Esquimaux. They are returned from this year's harvest of souls. If only he could tell them where he's just been. At this distance, it is impossible to identify who they are, but the brother feels a kinship just the same. He jumps up on the poop deck, removes his hat, and waves it in great, vigorous arcs. This greeting is returned before the tug speeds up and the *North Star* is pulled ahead, leaving all sailing traffic behind.

The following morning, the *North Star* is towed on the ebb tide to the mouth of the Thames under bright, welcoming skies. Miertsching takes his place at the gunwale, gaping at the parade of orderly beauty going by. It is a vision—long dreamt of and desired—that now springs into sensual existence. Tidy houses separated by emerald fields, bordered by towering, autumnal trees. Fragrant hearth smoke leavened with industrious clatter of rail cars. Countless well-dressed men and women go about their various pursuits and pleasures. That world, so long lost, is returned with a vibrancy and colour that clear away faded memories, replacing them with a palette of endless textures and hues. That passage of ice now seems impossibly remote, unconnected, as removed from life and the living world as are the planets or stars. Here, the brother observes, there are no muskoxen or caribou, whose flesh and blood sustained him in his hours of need. Here, he sees only "good," domesticated beasts that have found a useful place in man's ordered civilization, grazing a landscape over which man has clear domain. Returned from the icy wastes to this rich and generous place, his sense of gratitude overwhelms him.

"This is Woolwich," a sailor calls out.

And then the anchor is set.

The Investigators remain held onboard for yet another day and night. That longing to at last step ashore is prolonged further still. This journey, this life, seems but a series of frustrations endured.

A clutch of eager women and children can be seen on the dock below—smartly attired, faces upturned, awaiting the chance to welcome home husbands, fathers, sons. Word of the *North Star*'s approach was circulating long before her arrival. Expectations are high. But as the ship's captain moves among them, piteous scenes unfold as the widows and children of those men left behind are given the loathsome news. This heartbreaking view is one the brother wishes he could scrub from his eyes and memory.

At last, word comes down that the Investigators are to be transferred to the adjacent warship *Waterloo* to await further orders. Before they leave, a letter is read aloud from no less than the Commodore of the Admiralty, praising the Investigators for their exemplary conduct and service in the far-off fields of ice. News of their endurance and discovery, courtesy of Second Lieutenant Samuel Cresswell, has preceded them by a year. After this, Johann Miertsching—a civilian and foreigner—receives the gift of four days' leave. His comrades, however, cannot be released until an official inquiry is made into the loss of Her Majesty's ships.

Moments later, the brother finds himself caught up in the rush, pulled through town toward the railway station, toward London and the brethren, in the same patched and threadbare garb he received the day he staggered aboard the *Resolute* seventeen months ago. Along the way, he wanders into a haberdashery and orders up a suit of clothes. Half an hour later, he emerges into the light transformed. He proceeds down the street with travellers and townsfolk, turned out like any proper gentleman with people to meet and a schedule to keep.

On his return to the pier at Sheerness, October 12, Miertsching encounters numerous Russian prisoners from the recently declared war in Crimea. His comrades, still held awaiting trial aboard the *Waterloo*, greet him warmly. Together, they sit down to lunch and keep company for a few pleasant hours. When it is time to go, many of them escort the brother to the dock, where he will catch his boat back to London.

That first year in their midst was a difficult one, but he recalls the growing care, friendship, and even brotherly love he received as the months and trials multiplied. Now, with their damaged hands and feet, it is clear that some of them will no longer be able to ply their trades. Others, never truly recovered from scurvy and starvation, hope for a complete recovery in the care and comforts of home. The brother concludes that it is by God's grace alone that all but five have escaped with their lives. He considers the men's faces, battered by frostbite and exposure. Several of them now shed tears without reservation or shame. And when the moment comes to say goodbye, he too feels the pain.

On days such as these, a man is liable to take stock. While it is true that they failed in their mission to find and rescue

Franklin and his men, they succeeded in realizing an ancient dream—a Northwest Passage connecting the Atlantic and Pacific oceans. They have also become the first men to trace a line around the Americas. But as with so many dreams, the brother concludes, the realization of the Northwest Passage will not change a thing. It is plain to all that the treacherous, impassable route they discovered holds no significance—it is useless as a shipping lane. And, although he is neither a navigator nor a scientist, he suspects it will remain this way until some future age, when the climate is somehow utterly transformed and the thick, endless, shifting fields of ice are no more.

The following week, a court martial is convened aboard the *Waterloo* in the matter of the abandonment of five of Her Majesty's ships to the frozen sea. After a brief and perfunctory review of the facts, Robert McClure, captain of the *Investigator*, is not only honourably acquitted, he is given a commendation for gallantry and zeal exhibited, privations and hardships endured, and good services rendered to Great Britain—not least of which is his monumental discovery. Upon returning McClure's sword of command at the conclusion of his trial, Admiral George Gordon remarks, "The court are of the opinion that your conduct through your arduous exertions has been most meritorious and praiseworthy." Captains Henry Kellett and Sir Edward Belcher are also acquitted. Kellett, who maintains that he could have brought both the *Resolute* and *Intrepid* home, was reluctantly following orders. Belcher's case is another matter. Having ordered the abandonment of four ships under his ultimate command, his reputation is beyond repair. His sword is handed back in ringing silence.

On October 24, having secured both their release and pay, a committee of ten members of the *Investigator*'s crew immediately takes up a collection and raise more than £70. They invite Alexander Armstrong and his fellow officers to meet them in the pub of a nearby hotel—before the ship's company scatters across the British Isles and the sundry seas. When the party is comfortably assembled, a seaman stands. On behalf of the crew, he publicly thanks the surgeon for preserving them during their long ordeal. With genuine respect and affection, they present him with a fine gold watch and chain, along with a letter, signed by the men, declaring their gratitude for his unwavering service, attention, and care. Accepting the men's heartfelt gift and praise, Armstrong is moved to offer a simple, elegant reply. When the surgeon sits down, the sailor turns to Lieutenant Bedford Pim, and then speaks for everyone.

"If it had not been for you, sir, many of us now present would never have seen Old England again. All of us look upon you as a deliverer, and we shall never forget the joy we felt when you reached us."

In reply, Pim congratulates the men on having narrowly escaped the fate of Franklin.

This private event garners a small mention in the October 28 edition of the *Illustrated London News*. This notice is completely overshadowed by the announcement of the recent discovery of relics from the Franklin Expedition by the seasoned Arctic explorer Dr. John Rae. His report sprawls over almost an entire page. Ironically, Rae—in the employ of the Hudson's Bay Company—had not been searching for Franklin at the time, but trying to solve part of the puzzle of North America's Arctic coastline.

Rae learned from a group of Esquimaux that, in the spring of 1850, approximately forty white men were seen hauling a boat over the snow in the area of the Back River estuary on the mainland south of King William Island. They had no interpreter, but through signs were able to communicate that their ships had been crushed in the ice and that they were trying to make their way to Hudson's Bay territories. The Esquimaux sold them a seal, but the white men soon starved. From the Esquimaux, Rae purchased numerous relics having once belonged to Franklin and his men, including silver forks and spoons bearing officers' crests and a plate emblazoned with Franklin's name. He interviewed numerous individuals, testing the veracity of their claim. His report, published by the Admiralty while he was still at sea, also relayed the Esquimaux account that the starving men had turned to cannibalism in their final days. Bodies were reportedly seen in tents, under boats, scattered where they fell. Some appeared to have been butchered; human flesh was found in their kettles.

Lady Jane Franklin has long been agitating for and even financing rescue missions. She responds with a passionate attack on Rae's report, reputation, and character. After all, he trusted, lived with, and even dressed in the style of the savage Esquimaux—not civilized, British explorers like her noble husband. Regardless, Rae's evidence of Franklin's fate will earn the £10,000 parliamentary reward for himself and his men.

The day this edition of the *News* is being sold on the streets of London, Miertsching cashes his cheque for services rendered to the Royal Navy: £733. Within forty-eight hours, he receives his second written offer from the Admiralty for employment as interpreter aboard a British ship to be sent in search of more relics of the Franklin Expedition. Lady Franklin adds her urgent voice to the request for the brother's services, along with the promise of additional funds. He answers with an emphatic no, and the firm resolution to never again step foot in the Arctic.

Interest in the sensational revelations contained in Rae's report give way to accounts from the front lines of the Crimean War, which brings to an end half a century of peace between the major European powers. The day after Armstrong received his gold watch, the disastrous Charge of the Light Brigade took place against Russian forces in the Battle of Balaclava. By year's end, Lord Alfred Tennyson's poem will help refocus the passions and imagination of the nation. For both the Admiralty and the general public, the taste for Arctic adventures wanes. A new generation of young men is being called to another glorious duty and sacrifice with the promise that a grateful nation will never forget their names.

24.

RESTITUTION

Among veterans of the Arctic campaign, there has been a growing sense of impatience. Nearly eight months past the Investigators' return, a Select Committee of the House of Commons finally meets to discuss the matter of the £10,000 reward for the discovery of the Northwest Passage. In light of Rae's report, their duty is not so straightforward. Some members of the Admiralty are of the opinion that Franklin or his men may have discovered a Northwest Passage of their own, although they did not live to traverse it. However, the committee is compelled to come to a conclusion before the full extent of Franklin's fate is known.

Captains Henry Kellett and Richard Collinson each lay claim to part of both the honour and reward. Collinson has recently returned on the *Enterprise*. Instead of sailing through the chain of Aleutian Islands, he took the long way around its western end, costing him precious time. After reaching the ice nine days after the *Investigator*, he was forced to retreat and chose to winter in Hong Kong. When he returned to the Arctic in 1851, he finally succeeded in penetrating the ice and entered the Prince of Wales Strait, the first passage discovered by Robert McClure the previous year. Collinson wintered there while the *Investigator* was at the north end of Banks Island, beset in Mercy Bay. As McClure's commander, he claims a share in the discovery of the Northwest Passage.

Kellett's claim, by far the stronger, is based on having rescued the starving men of the *Investigator*. Despite the fact that on several occasions, both privately and publicly, McClure has declared Kellett his "preserver," he now sees no reason to share any of the credit. Remarkably, McClure responds that he could have saved his men and ship without assistance. It was Kellett who ordered him to abandon ship. Notwithstanding all the evidence given, notwithstanding the fact that five of his men perished even with Kellett's help, McClure contends that he could—as planned—have sent his invalids across the ice to safety and sailed the *Investigator* home with the loss of only four lives.

Armstrong, at sea with the Baltic Squadron, is unable to testify.

William Domville is called before the committee to report on the physical condition of the men aboard the *Investigator* upon his inspection of May 20, 1853. He admits that if help had not arrived, most would have perished. However, he obscures and underplays the severity of the medical emergency he encountered in Mercy Bay. When presented with McClure's claim—that he could have saved all but four of his men—Domville's response is unclear and evasive. When pressed to judge whether *any* of the Investigators could have embarked on McClure's planned journeys, Domville replies, "Some might have gone."

The day after Domville's testimony, McClure returns to deliver a two-sentence statement to the committee: "Perhaps I may be allowed to say a word with respect to Dr. Domville. When I went over with my men to the *Resolute,* they were immediately placed under Domville's charge, and he paid the greatest possible attention to them in their enfeebled state; and I think that some tribute is due to Dr. Domville for his great attention to them."

Not once does the name Alexander Armstrong pass his lips.

Kellett and the men of the *Resolute,* who had sought, rescued, fed, clothed, sheltered, and nursed the Investigators for a year—Kellett, who generously shared his private cabin with McClure—are given no more than a polite commendation by the committee. This very public and deliberate insult against him and his men prompts Kellett to offer each member of his ship's company £50 out of his own pocket.

Acting to protect the memory and legacy of her lost husband, Lady Franklin intervenes to ensure his sacrifice—and those of his men—is not forgotten. After all, McClure's passage was found in the search for Franklin. And yet the fact remains: abandoned ship and near catastrophe aside, the ambitious, bold, and deceitful McClure actually pulled it off. The committee finds in his favour, but with a conditional claim. In addition to becoming the first to circumnavigate the Americas, McClure and the men of HMS *Investigator* did indeed discover and cross *a* Northwest Passage between the Atlantic and Pacific. They are awarded the £10,000 prize, with half going to the captain and half divided among his officers and crew. For McClure, a knighthood is in the offing. It appears his place in the history of polar exploration is at last secure. From now on, whenever anyone thinks of the Northwest Passage, they will undoubtedly remember the name Robert John Le Mesurier McClure.

Fifty days following the conclusion of the Commons committee—September 10, 1855—the captain of an American whaling vessel cuts his losses in Baffin Bay. It has not been a fruitful voyage. With heavy fog, wind, and ice blocking the most productive whaling grounds, James Buddington has turned south to make a safe retreat for his ship, the *George Henry*, and his twenty-five-member crew. Along the way, he spies a dark shape in the grey and white. From a distance of fifteen miles, it appears to be a ship—although certainly no whaling vessel. Regardless, he would welcome the chance to exchange news and intelligence with her captain. But conditions prove so foul it takes five days to manoeuvre through the ice. After approaching the ship as closely as possible, all attempts to hail her crew are met with silence. Continued shouts and signalling bring no movement, light, or response of any kind. Buddington orders four of his men to scramble over the ice and water to investigate. They soon find themselves standing onboard the ghost ship *Resolute*.

The Yankee whalers kick in the hatch and peer down into the shadows. They sense no signs of life stirring inside. A haunting stillness hangs in the air, but there is not a single corpse to be found. Instead, they discover open books, tables with cutlery and plates neatly arranged and set, half-filled glasses of wine, a game of cards half played. A captain's epaulettes are found draped over a chair. Everything remains as if the officers and crew suddenly, inexplicably vanished.

Overcoming their superstitions and fears, Buddington's men dress up as officers of the Royal Navy and engage in bouts of mock swordplay. They toast the vanished crew with the stock of wine left behind. They do not realize that this vessel, abandoned sixteen months ago, has travelled twelve hundred miles back through the Northwest Passage, or that this miraculous self-navigation dramatically underscores her absent captain's claim that he could have sailed her home.

With no small effort or risk, Buddington sails the ghost ship to his home port of New London, Connecticut. Along the way, he encounters another British ship and hands off Kellett's epaulets with the request that they be reunited with their owner. If that captain is still alive, he should be among the first to learn that his ship endures.

The salvaged *Resolute* will eventually be purchased by the United States government for $40,000. She will be refurbished, returned to England, and presented to Queen Victoria as a gesture of friendship and peace. When the ship is finally retired from service and broken up in 1879, the British government will order that timbers from the vessel be used to build a sturdy desk. The following year, the desk will be shipped back across the Atlantic to Washington, D.C., and presented to President Rutherford B. Hayes. Known as the Resolute Desk, it will reside in the White House and be used in either the Oval Office or private study of nearly every subsequent president of the United States. It will live on as a bond between seafaring nations, recalling the perils and mysteries of polar navigation.

Return of the *Resolute* aside, the Arctic ice keeps far more secrets than it is willing to release. Before the close of 1855, a report will appear in the Parliamentary "Blue Book" on Polar Expeditions by the last man known to have seen the 422-ton ship abandoned by Robert McClure. His account will stand in the permanent record as the final word on the fate of HMS *Investigator*.

25.

DISSOLUTION

May 5, 1854

By the time he sights the derelict beset in Mercy Bay, he has been plodding across the ice for a month. Commander Frederick J. Krabbé, along with a sledge party of six men, has been dispatched from the *Intrepid* under orders from Captain Henry Kellett. They are to ascertain the condition and situation of the ship and retrieve the journals left behind by Robert McClure eleven months ago. As Krabbé and his men approach from the north and east across the frozen strait, the ragged remains of the ship's ensign and pennant are heard and seen snapping above the deck. A drift of snow has accumulated on the north side of the vessel in such a way that when the men arrive at half past ten in the twilit night, they are able to simply step aboard.

The *Investigator* points north and west. She stands 142 paces from shore in eleven fathoms of water. A cable hangs slack under her bow. She leans forward slightly and lists about 10 degrees to starboard. Although there are no signs of undue pressure from the ice, oakum is seen hanging from her seams. Having taken care to inspect the condition of the ice on his way into Mercy Bay, Krabbé is convinced that it never broke up during the summer of 1853. It was indeed fortunate that the Investigators abandoned ship when they did. If his fellow officer, Lieutenant Bedford Pim, had never arrived, the scene greeting him this day would be of a wholly different character.

As the men descend below deck, everything appears in good order and free of frost. Farther down the lower decks, however, the beams and planks overhead are covered with great accumulations of ice. In the holds, the men stand on a solid floor of frozen sea—proof the ship leaked considerably over the preceding summer. Their initial inspection complete, Krabbé and his weary men choose vacant cabins and berths, spread blankets, and settle down for the night.

Following breakfast, they begin an inventory of the useful supplies still onboard. As it is beyond doubt that the ship will take on more water this coming season, the decision is made to shift all remaining provisions to the depot already established on shore. Over the next four days, the men pack and off-load the following:

1 cask salt beef
1 harness cask salt pork
1,500 lbs. preserved meat
3 casks flour
12 casks biscuit
1 cask suet
1 cask sugar
50 lbs. tea
5 cases tobacco
230 yards flannel
80 lbs. wool serge
100 pairs mitts
15 jackets
sailcloth, housing cloth, and tents.

These provisions are loaded onto the sledge, hauled to the beach, and piled beside the original depot left by the Investigators for the relief of mariners in need, namely Sir John Franklin or Captain Richard Collinson. A stock of wine, spirits, coal, and a further 3,300 pounds of preserved meat cannot be moved, as they are frozen aboard in a solid mass of ice.

This work done, Krabbé and his men take the next two days to select and pack provisions for their return across the strait. To their sledge, they add precious medicines and thermometers. From among the vast collection of stones and cinders, pressed flowers, herbs, and rigid birds found onboard the vessel, they choose a small number of specimens. After a thorough search, however, not a single journal is found.

At six o'clock on the evening of May 11, they secure the hatches and take up their harnesses. Krabbé composes a record of their visit and deposits it in the cairn at Point Providence. They leave confident that the *Investigator* will soon sink at her anchors and ultimately find repose on the bed of Mercy Bay.

Neither Franklin nor Collinson will ever reach this place. Collinson will return to England with the *Enterprise* and lay his claim against McClure. The fate of Franklin and his men will be further revealed by Francis Leopold McClintock, former commander of the *Intrepid*. In 1859, McClintock will sail the yacht *Fox* to the Arctic on a private expedition, financed by Lady Franklin, to investigate Dr. John Rae's claims. He will meet groups of Esquimaux who confirm accounts of ships being crushed in the ice near King William Island, 550 miles to the southwest of Mercy Bay. He will find a note left by a member of the expedition, dated April 25, 1848, stating that Franklin had died in June the previous year and that both the *Erebus* and *Terror* had been abandoned to the ice. Not unlike the plans Robert McClure had for the weakest members of his ship's company, a group of Franklin's men were about to begin a desperate journey south by sledge and boat.

At King William Island, McClintock and his men will discover a 650-pound oak-and-iron sledge laden with an astonishingly large assortment of equipment, clothing, books, sheet lead, cutlery, and a twenty-eight-foot boat weighing upwards of eight hundred pounds. Inside the boat he will find two human skeletons. McClintock's expedition will establish the probability that a few of Franklin's men travelled far enough to discover the last link of a Northwest Passage through what will become known as Franklin Strait—predating by a least two years the discoveries of Robert McClure. What is considered a certainty is that none of them traversed or survived it. Regardless, a statue will be raised in Spilsby, Lincolnshire—Franklin's hometown—with the following inscription:

SIR JOHN FRANKLIN—

DISCOVERER OF THE NORTH WEST PASSAGE.

Statues in both London and Australia will bear inscriptions making similar claims. A public so long enthralled by—and invested in—Franklin's saga will make him the sentimental favourite over the ungrateful survivor Robert McClure. Regardless, the unprecedented search for Franklin will ultimately reveal that there are in fact several possible routes for a Northwest Passage—all either partially or perpetually choked with ice.

The depot left at Mercy Bay will not go unnoticed by resident and migratory birds. Arctic foxes, wolves, and polar bears are sure to make the discovery. While the day, time, and name of the first person to find the exotic cache will forever remain a mystery, word will spread among the people who will become known as Copper Inuit for their use of the native ore.

Groups will begin to make annual treks north from traditional hunting grounds to a land once well used by their ancestors. Tales of these journeys will be handed down. They will travel through the heart of the island, along the river and valley that leads toward Mercy Bay. Their path will take them across some of the richest forage in the western Arctic, and they will avail themselves of abundant muskoxen and caribou encountered along the way. Following years of overhunting, these stocks will begin to decline.

For generations of Inuit, the metal, rope, wood, clothing, fabric, and food stashed on the north coast of the island will prove irresistible. This unprecedented trove will alter their technology, material wealth, and ability to trade. It will change the course of their lives. Those items found already useful or attractive will be put to immediate use. Others will be repurposed or reshaped. Those things deemed strange or unnecessary will be considered, then tossed aside.

The cryosphere will expand and contract fifty-one times. The polar ice pack will make twelve complete rotations around the Arctic Basin. Then, in 1905, a Norwegian explorer will pass Banks Island's southern shore on what is destined to become the first successful voyage through the Northwest Passage in a single vessel—the slight, seventy-foot, forty-seven-ton sloop *Gjøa*. Roald Amundsen and a crew of six will have travelled the southerly route, via Victoria Strait and Coronation Gulf, through treacherous channels as shallow as three feet. Their two-year passage will prove impossible for larger vessels.

Four summers later, a sledge crew led by Second Officer O.J. Morin of the Canadian Coast Guard Ship *Arctic* will appear at the mouth of Mercy Bay. He will have arrived as Pim and Krabbé before him—having crossed the impenetrable ice from Melville Island over what has become known as McClure Strait. At Providence Point, Morin will discover that the cairn has been demolished. Pushing deeper into the frozen bay, he and his men will make a thorough search but find no trace of HMS *Investigator*. The only remains of the depot will be piles of coal, fragments of sail and rope, a scattering of barrel staves. Even the graves will be gone. No monument or written record will be found. Nearly all evidence of ambition, suffering, and endurance will have been picked clean, wiped away. Before returning to his ship at Winter Harbour, Morin will rebuild the cairn. Using the barrel of his rifle as a flagstaff, he will raise a red ensign and take formal possession of Banks Island (ceded from Great Britain in 1880) for the Dominion of Canada. He will neatly fold the flag with an account of his visit and tuck it into the cairn.

A generation will pass before another vessel navigates the Northwest Passage: the eighty-ton iron-clad schooner *St. Roch*. Her captain, Henry Larsen, will be on another Canadian sovereignty cruise—this time, in a world at war.

Between 1942 and 1944, the *St. Roch* will be the first to traverse the passage in both directions, retracing the treacherous southern route of the *Gjøa* (west to east), then returning through Parry Channel to Melville Island and down the first of McClure's routes—along the eastern shore of Banks Island through the Prince of Wales Strait. Following the war, an unknown and increasing number of submarines will pass under the ice, through the Northwest Passage, and throughout the Arctic Basin.

A century after the United States purchased Russian America and renamed the territory "Alaska," the largest oil field on the continent will be discovered on the Arctic coast at Prudhoe Bay in 1968. The following year, the U.S. supertanker *Exxon Manhattan*, escorted by the Canadian Coast Guard icebreaker *John A. Macdonald*, will become the first commercial vessel to navigate the Northwest Passage. Undertaken to test the feasibility of tanker traffic through Arctic ice, this voyage will pass from east to west by way of the far more important northern route, through Parry Channel, until heavy ice at McClure Strait forces them down and around Banks Island through the Prince of Wales Strait. Upon arrival at Prudhoe Bay, a single, token barrel of oil will be loaded aboard the supertanker. Six years later, construction of the overland Trans-Alaska Pipeline System will commence.

Reliable satellite data of Arctic sea ice coverage will become available in 1979. For a time, the expansion and contraction of first-year and multi-year ice will continue within a stable range. Then, as the twentieth century draws to a close, a series of thaw-freeze events will take place on Banks Island and throughout the Arctic Archipelago, preventing muskoxen and caribou from reaching their forage. Starving herds of both species will be found wandering out on the ice, miles from shore, in a desperate search for sustenance. Catastrophic die-offs will occur. These same events will trigger a decline in the lemming population, affecting falcons, snowy owls, Arctic foxes, and Arctic wolves—the last wolf subspecies still found throughout its original range. Due to the decrease in spring snow cover, an exploding snow goose population will begin to pose a threat to their fragile breeding grounds.

A thinning of the multi-year polar pack and a decline in the extent of summer sea ice will become pronounced. Ice-dependent species, such as walrus, ringed seals, and polar bears, will face a shrinking and fragmented habitat. Permafrost will begin to melt, resulting in landslides and cliffs slumping into the sea. On Banks Island, as throughout much of the Arctic, travel across land, ice, and open water by Inuit hunters will become more difficult and dangerous as weather patterns and seasons become unpredictable and temperatures rise.

Having narrowly escaped extinction, bowhead whales will undergo a long, slow recovery following a commercial whaling ban. The Inuit, however, will continue to hunt them off the north coast of Alaska as they have for over a thousand years. In the 1990s, several antique harpoon fragments will be discovered embedded in the flesh of freshly killed whales. This evidence—combined with biological indicators found within the whales' eyes—will lead scientists to conclude that bowhead whales are likely the longest-lived mammals on Earth. One specimen, killed in 1999, will be determined to have lived for 211 years. It had been navigating the polar seas for over sixty years before the *Investigator* set sail on her final voyage.

By the summer of 2007, sea ice coverage will be 39 percent below the long-term average observed between 1979 and 2000. When ship and aircraft observations prior to satellite data are taken into account, the ice will be found to have shrunk by as much as half over the preceding fifty years. On September 16 of that year, satellite data will also confirm that the main, deep, northern channel of the Northwest Passage—Parry Channel to McClure Strait—will not only be open for the first time in human memory, it will be ice-free. For five weeks, fortified icebreakers or small, specialized craft will not be necessary for plying the route. Standard ocean-going tankers and containerships will be able to pass with ease.

Where Inuit elders, oceanographers, biologists, and climatologists see rising cause for alarm, shipping syndicates will perceive opportunity. They will draw up plans for new vessels to navigate the thinning or nonexistent ice. But just as the old dream of an open Northwest Passage finally comes true, so will a dawning realization that the route will likely be abandoned almost as soon as it comes into regular use.

Stunned by the pace of Arctic melt and warming, an international consortium of scientists will revise earlier estimates of sea ice decline. They will begin to foresee the complete disappearance of multi-year sea ice within their own lifetimes. They will predict that by the summer of 2030 the Arctic will be seasonally ice-free. When this occurs, mariners will not bother with the longer route through the Arctic Archipelago. They will load their ships with oil and merchandise and sail directly across the top of the world through the Open Polar Sea.

AUTHOR'S NOTE

IT IS BOTH A GREAT PRIVILEGE and responsibility to resurrect this critical episode in the lives of men who lived more than a century and a half ago. Their rich and varied accounts have served as remarkable guides.

In 1854, based on interviews with Johann August Miertsching, Daniel Benham published *Sketch of the Life of Jan August Miertsching, Interpreter of the Esquimaux Language to the Arctic Expedition on Board HMS* Investigator, *Captain McClure, 1850, 1851, 1852, 1853*. Published by Miertsching's friend through a small Moravian mission press in London, this booklet offers highlights of the expedition and a flattering portrayal of Captain Robert McClure.

In 1856, Captain Sherard Osborn, Royal Navy, published *The Discovery of the North-West Passage by HMS "Investigator," Capt. R. McClure, 1850, 1851, 1852, 1853, 1854*. This was the authorized and heavily edited version of McClure's own journal as captain of HMS *Investigator*. A deliberate whitewash of McClure's mistakes and the suffering of the ship's company, it is also marked by an excess of Victorian sentiment and pride.

The following year, based on his own journal, Alexander Armstrong published *A Personal Narrative of the Discovery of the Northwest Passage; With Numerous Incidents of Travel and Adventure During Nearly Five Years' Continuous Service in the Arctic Regions While in Search of the Expedition under Sir John Franklin.* Armstrong's book is a serious and detailed report of the expedition, as well as an often-scathing indictment of McClure's judgment and decisions. Undoubtedly, at least part of the reason for its publication was an attempt to correct the public record and settle scores.

Assistant Surgeon Henry Piers kept a personal journal that has never been published. It confirms most major events mentioned in the other accounts, while offering colourful details of daily life and encounters with wildlife. The journal abruptly ends on March 31, 1852, omitting the harshest year before the rescue of the Investigators. The whereabouts of the missing portion—if ever completed—remains unknown.

Before abandoning ship, McClure ordered that all journals be surrendered to him. Likely, there were many. It seems certain this was done to prevent criticism of his leadership or decisions, to avoid graphic accounts of suffering and privations, to allow him to dictate the official version of their story and reap the rewards with publishers. In the mid-nineteenth century, the market for narratives of polar discovery and adventure was robust. How Armstrong and Piers were able to rescue their journals is unclear. Perhaps an arrangement had been made with William Domville to have them brought to the *Resolute* under Captain Henry Kellett's orders. They would have provided strong evidence of the desperate physical conditions aboard the *Investigator*. Perhaps the two surgeons never complied with McClure's order, or they were able to find and retrieve their confiscated journals and smuggle them across the strait.

In 1967, the full German text of Brother Johann August Miertsching's personal journal was translated into English by L.H. Neatby and published for the first time as *Frozen Ships: The Arctic Diary of Johann Miertsching, 1850–1854*. Never intended for publication, recounted from memory (with McClure's own journal as a guide), told by a man for whom Royal Navy society and traditions were foreign, it provides a sensitive outsider's account of the expedition.

Additional sources helped shed light on the lives of the Investigators and the Inuit in the western Arctic during the nineteenth century. Those I found most valuable are included in the bibliography.

It is difficult to know with certainty what was actually said over 150 years ago. In the case of the Investigators, however, some strong evidence endures. When, in the narrative, I quote dialogue, it is taken either from direct quotes found within the various accounts or reconstructed from one person paraphrasing another. In both cases, the words from source documents are changed as little as possible, and only for the sake of sense, continuity, or grammar.

Never before have these various accounts been considered together. Among them are inconsistencies, contradictions, and omissions. In re-creating this shared adventure, I have included nothing I know to be untrue. Any mistakes are my own.

SELECTED BIBLIOGRAPHY

Armstrong, Alexander. *A Personal Narrative of the Discovery of the Northwest Passage; With Numerous Incidents of Travel and Adventure During nearly Five Years' Continuous Service in the Arctic Regions While in Search of the Expedition Under Sir John Franklin*. London: Hurst and Backett, 1857.

Ashford, Graham and Castleden, Jennifer. "Inuit Observations on Climate Change Final Report." International Institute for Sustainable Development, June 2001.

Barr, William. *Arctic Hell-Ship: the Voyage of the HMS Enterprise 1850-1855*. Edmonton: The University of Alberta Press, 2007.

Bartholomew, Michael. "James Lind and Scurvy: A Revaluation," *Journal for Maritime Research*, Jan. 2002.

Benham, Daniel. *Sketch of the Life of Jan August Miertsching, Interpreter of the Esquimaux Language to the Arctic Expedition on board HMS Investigator, Captain McClure,*

1850, 1851, 1852, 1853. London: William Mallalieu and Co., 1854.

Bernier, J.E. *Report on the Dominion Government Expedition to Arctic Islands and the Hudson Strait On Board the D.G.S. "Arctic."* Ottawa: Canada Dept. of Marine and Fisheries, Govt. Printing Bureau, 1910.

Berton, Pierre. *The Arctic Grail: The Quest for the North West Passage and the North Pole, 1818-1909*. Toronto: McClelland and Stewart, 1988.

The Birds of North American Online. Cornell Lab of Ornithology, (http://bna.birds.cornell.edu/bna).

Brown, John. *The North-West Passage, and the Plans for the Search for Sir John Franklin: A Review*. London: E. Stanford, 1858.

Circumpolar Flaw Lead System Study. University of Manitoba, (www.ipy-cfl.ca).

Condon, Richard G. with Ogina, Julia and the Holman elders. *The Northern Copper Inuit: A History*. Norman: University of Oklahoma Press, 1996.

De Bray, Emile (trans. and ed. by Barr, William). *A Frenchman in Search of Franklin: De Bray's Arctic Journal, 1852-1854*. Toronto: University of Toronto Press, 1992.

English, John, ed. Dictionary of Canadian Biography Online, (www.biographi.ca/index-e.html).

Grayhound Information Services, The Natural and Cultural Resources of Aulavik National Park, (report prepared for Parks Canada). March, 1997.

Harrod, Dominick. *War, Ice & Piracy: The Remarkable Career of a Victorian Sailor, The Journals and Letters of Samuel Gurney Cresswell*. London: Chatham Publishing, 2000.

The Illustrated London News, October 28, 1854.

Inuvialuit Pitquisiit: The Culture of the Inuvialuit. Published by the Government of the Northwest Territories (Education), 1991.

Isserman, Betty. *Sinews of Survival: The Living Legacy of Inuit Clothing*. Vancouver: UBC Press, 1997.

Krupnik, Igor and Jolly, Dyanna, eds. *The Earth is Faster Now: Indigenous Observations of Arctic Environmental Change*. Fairbanks, Alaska: Arctic Research Consortium of the United States, 2002.

Lind, James. *A Treatise on the Scurvy*. London: A. Millar, 1753.

Lopez, Barry. *Arctic Dreams: Imagination and Desire in a Northern Landscape*. New York: Scribner, 1986.

Marriott, Bernadette M. and Carlson, Sydne J., eds. *Nutritional Needs in Cold and in High-Altitude Environments, Applications for Military Personnel in Field Operations*. (Committee on Military Nutrition Research, Food and Nutrition Board, Institute of Medicine.) Washington: National Academy Press, 1996.

McClure, Robert. Correspondence from Commander Robert McClure to the Secretary of the Admiralty from HMS *Investigator* (at sea). July 20, 1850.

———. Unpublished correspondence between Robert McClure and Sir George Back.

McDougall, George F. *The Eventful Voyage of H.M. Discovery Ship "Resolute" to the Arctic Regions in Search of Sir John Franklin and the Missing Crews of H.M. Discovery Ships "Erebus" and "Terror," 1852, 1853, 1854*. London: Longman, Brown, Green, Longmans, & Roberts, 1857.

National Oceanic and Atmospheric Administration (NOAA). United States Department of Commerce, (www.noaa.gov).

National Snow and Ice Data Center. University of Colorado, Boulder (http://nsidc.org/).

Neatby, L.H. *Frozen Ships, the Arctic Diary of Johann Miertsching 1850-1854*. Toronto: Macmillan, 1967.

Nikels, S., et al., eds. *Unikkaaqatigiit: Perspectives from Inuit Canada (Putting the Human Face on Climate Change)*. Laval: Inuit Tapiriit Kanatami, Nsasivvik Centre for Inuit Health and Changing Environments at Université Laval and the Ajunnginiq Centre at the National Aboriginal Health Organization, 2005.

Osborn, Sherard, ed. *The Discovery of the North-West Passage by HMS "Investigator," Capt. R. McClure, 1850, 1851, 1852, 853, 1854*. (From the logs and journals of Capt. Robert Le M. McClure.) Second edition. London: Longman, Brown, Green, Longmans, & Roberts, 1857.

Oswalt, Wendell. *Eskimos and Explorers*. Lincoln: University of Nebraska Press, 1999.

Paine, Lincoln. *Ships of Discovery and Exploration*. London: Houghton Mifflin, 2000.

Pielou, E.C. *A Naturalist's Guide to the Arctic*. Chicago: University of Chicago Press, 1994.

Piers, Henry. "The Diary of Assistant Surgeon Henry Piers." Dec. 1849–March 1853. Unpublished.

"Report from the Select Committee on Arctic Expedition; Together with the Proceedings of the Committee, Minutes of Evidence, and Appendix." London: House of Commons. July, 1855.

Ross, Sir John. *Narrative of a Second Voyage in Search of a North-West Passage*. London: A.W. Webster, 1835.

van Everdingen, Robert O., compiled by. "Proceedings: 14th Inuit Studies Conference, 11-15, August 2004."

The Arctic Institute of North America, University of Calgary.

Sandler, Martin. *Resolute: The Epic Search for the Northwest Passage and John Franklin, and the Discovery of the Queen's Ghost Ship*. New York: Sterling, 2006.

Savours, Ann. *The Search for the North West Passage*. New York: St. Martin's Press, 1999.

Thomas, David. *Frozen Oceans: The Floating World of Pack Ice*. Buffalo: Firefly, 2004.

Watt, J. et al., eds. *Starving Sailors: The Influence of Nutrition Upon Naval and Maritime History*. Bristol: John Wright & Sons Ltd. (on behalf of the trustees of the National Maritime Museum), 1981.

ACKNOWLEDGEMENTS

THIS BOOK EXISTS BECAUSE Tim Rostron, Senior Editor at Doubleday Canada, took a risk. His faith and patience allowed me to find my own way to this story; his insight and wisdom helped me sharpen its focus. I am indebted to the entire team at Doubleday Canada, especially Susan Burns, who oversaw the book's production; Lloyd Davis, who carefully copyedited the text; Terri Nimmo, who designed its evocative cover; Paul Dotey, who created the excellent maps; Dory Carr-Harris, who helped put it all together; and Adria Iwasutiak, who has worked so hard to make it known.

Over the course of my research, I was the beneficiary of much goodwill and assistance. Numerous people generously offered their time, stories, resources, food, shelter, and company. On Banks Island, I am especially indebted to Edith Haogak, John Keogak, and David Haogak, Site Manager, Aulavik National Park. Onboard the CCGS *Amundsen*: Captain Stéphane Julien, Dan Leitch, Circumpolar Flaw Lead (CFL) Study Coordinator, and David Barber, Principal

Investigator (CFL), Canada Research Chair in Arctic System Science, and Director of the Centre for Earth Observation Science at the University of Manitoba. At Parks Canada in Inuvik: Melinda Gillis, Heritage Communications Officer, and Pat Dunn, Manager, Communications and Visitor Services.

Countless research requests and inquiries were patiently handled by the staff at the National Maritime Museum's Caird Library in Greenwich, the Scott Polar Research Institute at the University of Cambridge, the Royal Geographical Society in London, and the Vancouver Maritime Museum.

I am grateful to my agent, Georges Borchardt, for his dedication, enthusiasm, and advice. Thanks are also due to Angele Watrin Prodaehl, for her keen eyes, and to the estimable Vancouver FCC—Deborah Campbell, Charles Montgomery, J.B. MacKinnon, Alisa Smith, Chris Tenove, and John Vaillant—for their friendship and good counsel.

One person was already behind this venture even before I knew where I was headed: my first reader and beloved wife, Lily Harned.

My biggest debt, of course, is to the men I can never thank. At its heart, this book is a collaboration with the officers and crew of the *Investigator*, who, according to their characters, revealed their hopes and fears. They allowed me to experience an extraordinary time and place through their senses. In doing so, they have helped me view the natural world—and human nature—with greater clarity.

A NOTE ABOUT THE AUTHOR

BRIAN PAYTON has written for *The New York Times*, *Los Angeles Times*, *Chicago Tribune*, *Boston Globe*, and *The Globe and Mail*. He is the author of the novel *Hail Mary Corner* and the non-fiction narrative *Shadow of the Bear: Travels in Vanishing Wilderness*. Payton lives with his wife in Vancouver. Visit his website: www.brianpayton.com